IMAGES
of Aviation

ELLINGTON FIELD

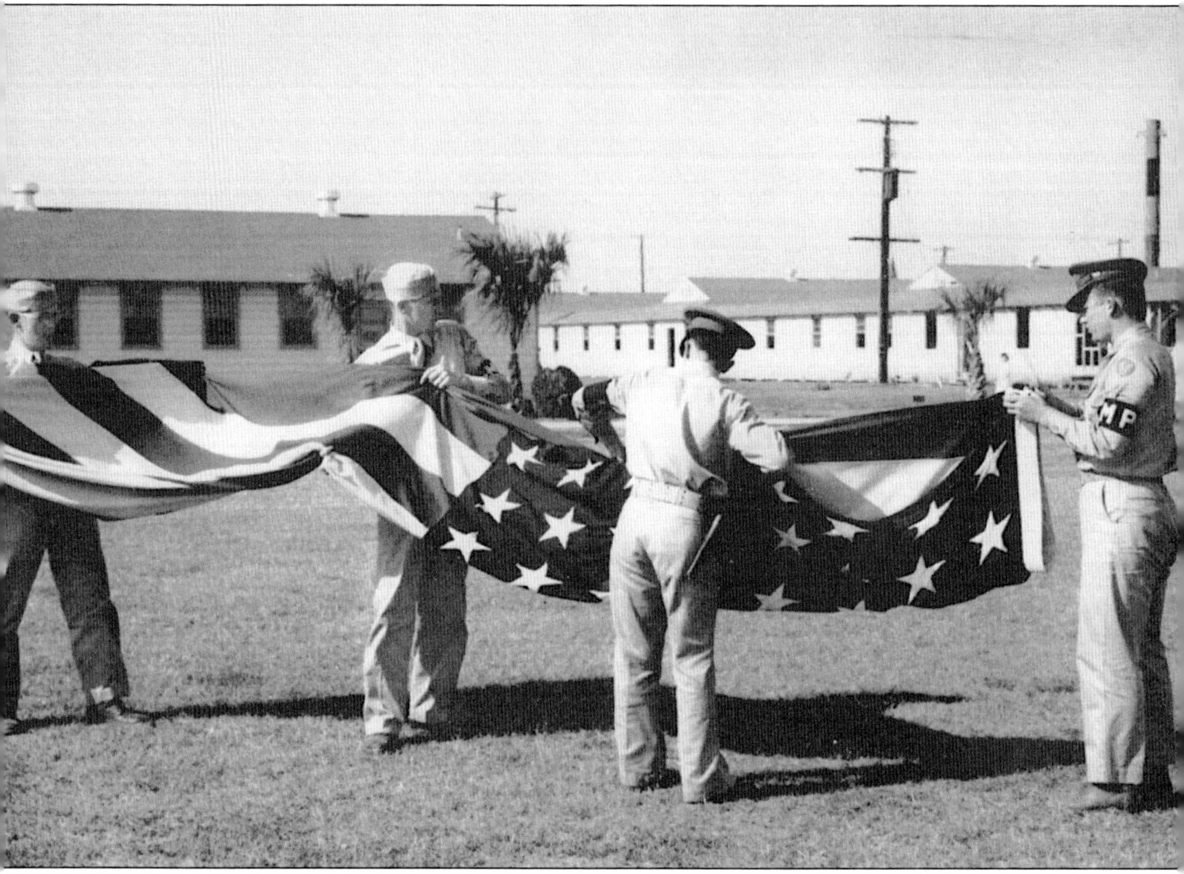

Military police retire the colors at Ellington Field at the end of the day during World War II. (Lone Star Flight Museum.)

On the Cover: Officers exit a C-45 VIP transport for a visit at Ellington Air Force Base in the early 1950s. During World War II, a different version of this aircraft, the AT-7, was used as a multi-engine trainer at then Ellington Field. (Lone Star Flight Museum.)

IMAGES
of Aviation

ELLINGTON FIELD

Mark Lardas

ARCADIA
PUBLISHING

ISBN 978-1-4671-0480-7

Published by Arcadia Publishing
Charleston, South Carolina

Printed in the United States of America

Library of Congress Control Number: 2019952523

For all general information, please contact Arcadia Publishing:
Telephone 843-853-2070
Fax 843-853-0044
E-mail sales@arcadiapublishing.com
For customer service and orders:
Toll-Free 1-888-313-2665

Visit us on the Internet at www.arcadiapublishing.com

To my nephew, John Lardas

Contents

ACKNOWLEDGMENTS

No book is the product of a single author. Many folks helped me as I wrote and assembled this book. I especially would like to note the following people for their assistance: my son William Lardas for accompanying me on my trips to the Lone Star Flight Museum and for doing the photography there; all the wonderful folks at the Lone Star Flight Museum, including, but not limited to, Stewart Bailey (curator) and Rachel Hermann (director of collections); Shawn Harrison, Chris Remy, Carl Brainerd, David Alexander, and Will Stager for offering photographs to use in the book (I did not use them all, but all were appreciated); Missy Bryant for lending me her copy of *From Jennies to Jets: The Story of the 111th Squadron*, a book her father wrote while with the 111th, and also for sharing her memories of her father and his experiences at Ellington; and Senior Master Sgt. Sean S. Cowher of the 147th Attack Wing History Office for his help with the history of the 111th Squadron and the 147th Attack Wing, to which the 111th belongs.

I would also like to thank Lindsey L. Givens and Caitrin Cunningham, my editors at Acadia Publishing, for their assistance.

The following abbreviations indicate the sources of the illustrations used in this volume:

147th Strike Wing History Office, Ellington Joint Reserve Base (147th)
author photograph (AP)
Defense Visual Information Distribution System (DVIDS)
Federal Aviation Administration (FAA)
Houston Airport System (HAS)
Library of Congress (LOC)
Lone Star Flight Museum (LSFM)
National Aeronautics and Space Administration (NASA)
National Museum of the US Air Force (NMUSAF)
photograph by William Lardas (WL)

INTRODUCTION

In 1913, aviation was 10 years old. Yet flying caught the imagination of Eric Ellington. Graduating third in the US Naval Academy class of 1909, he resigned his naval commission in 1911 to pursue flying. Accepting a second lieutenant's commission in the US Army in 1911, in 1912 he transferred to the Aeronautical Division, US Signal Corps. Within a year, he had pilot's wings and was a flight instructor.

In some alternate history, Ellington might have gone on to become America's leading fighter ace of World War I, win the first Bendix Trophy in 1931, command the mighty 8th Air Force in World War II, or become the first general of the Air Force. Instead, on November 24, 1913, during flight training with a student, his airplane crashed. Ellington and his student, Lt. Hugh Kelly, both died.

World War I started in 1914. Three years later, the United States joined the war. The military needed thousands of pilots, observers, bombardiers, and mechanics to operate and maintain aircraft for what had become the Aviation Section, US Signal Corps, in 1914. In 1917, Congress authorized construction of 32 air training camps. One, to be built on a swampy cow pasture southeast of Houston, was named for Eric Ellington.

Construction on Ellington Field started September 14, 1917. Work progressed swiftly. The first buildings, and more importantly the first runway, were complete by November. On November 29, the runway was used for the first time.

In 1917, cross-country flying was rare. New aircraft were crated up after an initial factory assembly, shipped to their duty airfield, and assembled there. This meant that at Ellington, initial emphasis was given to building workshops and hangars, then communal buildings such as the mess hall and hospital, and only then barracks and housing. For the first year, Ellington's garrison mostly lived under canvas, in soggy rows of five-man tents.

By November 1918, when World War I ended, Ellington was one of the most highly regarded training fields in the United States, with a reputation as the Army's best-equipped and most complete flying school. It housed over 250 aircraft and 5,000 troops. It provided training in flying, navigation, bombing, and radio. It pioneered night flying and created the nation's first air ambulance.

Ellington grew so quickly, satellite fields were built at Park Place and San Leon. These provided extra space for aircraft and destinations for student pilots. An airfield in far-off Katy, Texas (now inside Houston's Beltway), provided cross-country flying opportunities.

After the war ended, Ellington shrunk almost as quickly as it had grown. Park Place and San Leon closed in 1919, and 1,200 crated aircraft awaiting assembly were sold as surplus. In 1920, most of Ellington's garrison was shipped elsewhere. Only a skeleton crew remained at Ellington. The 1st Pursuit Squadron was stationed there in 1921–1922, but then transferred. Texas's National Guard 36th Division used Ellington for training from 1923 through 1927. Its aviation detachment, the 111th Squadron, was then stationed at Ellington until a fire destroyed many of Ellington's remaining buildings in 1927. The 111th then moved to Houston Municipal Airport, today's Houston Hobby Airport.

By 1930, Ellington was almost totally abandoned. The underground water pipes had been removed, as had the railroad tracks at the Ellington siding. Only the runway remained. From 1930 through 1940, the Army leased out the airfield for grazing cattle. It was once again cow pasture. A new world war revived Ellington. Japan attacked China in 1937. Italy conquered Albania in 1939. Germany and the Soviet Union invaded Poland later that year. Other nations joined in, and the wars merged and grew into World War II.

As during World War I, the United States was initially neutral. Unlike World War I, US preparations began well before the nation entered the war. In 1940, Congress allocated $7 million to rebuild and reopen Ellington. It was again to be used as a training field. Part of Matagorda Island was purchased as a bombing range for Ellington-based fliers. It showed how much aviation had matured since Katy was a distant destination from Ellington.

Ellington reopened in December 1940. Its first commander was Lt. Col. Walter Reid, who had taken advanced training at Ellington in 1918. Over the next four months, 165 buildings were completed: warehouses, barracks, mess halls, offices, chapels, and a hospital. Most were "temporary" wood frame structures, quickly thrown up for the duration. Only hangars and maintenance buildings had metal and concrete construction.

Electrical and telephone wires, plumbing and sanitation, and roads were installed. Concrete runways, taxiways, and aprons replaced their World War I–era grass, dirt, and packed-gravel counterparts, creating the runway layout still used today.

The first class of aviation cadets started training on October 6, 1941. They graduated on December 12, 1941, five days after Japan launched an aerial attack on Pearl Harbor and Oahu, four days after the United States declared War on Japan, and one day after Germany and the United States went to war with each other. That class graduated 50 pilots, 28 bombardiers, and 144 preflight cadets.

Flight training continued throughout World War II. Advanced bombing training was dropped and moved to less cloud-covered bases. Navigation was added. Emphasis shifted from primary to advanced pilot training. Between December 12, 1941, and September 2, 1945, over 15,000 pilots, navigators, gunners, radio operators, and preflight cadets graduated from Ellington's many training schools.

Ellington peaked in early 1944. By late in that year, aircrew surpluses and rifleman shortages decreased flight training's importance. After the war ended, Ellington Field was deactivated on March 15, 1946.

This time, the Cold War led to its reactivation. The Soviet Union, our World War II ally, occupied Eastern Europe and threatened Western Europe and the United States. Ellington reopened in 1948 as Ellington Air Force Base, part of the United States Air Force—formed in 1947 as an independent service replacing the Army Air Force.

The 111th returned to Ellington in 1948 as a Texas Air National Guard fighter-bomber squadron. Nationalized for service in 1950, it was sent to Korea. In 1949, Ellington reopened its training function. Between 1949 and 1956, the school trained over 19,000 navigators. It was the Air Force's only postwar air navigation school, teaching a skill critical to staffing the Air Force's expanding fleet of heavy bombers and multi-engine transports.

The 111th Squadron, now the 111th Fighter Interceptor Squadron, returned to Ellington in November 1956. The unit, with its parent 147th Group (and later wing) remained at Ellington thereafter. Ellington had a new mission: air defense. It became part of an early warning and air interception network protecting the upper Gulf Coast and its critical energy industry from enemy air attack. The threat seemed hypothetical in 1956. After 1960, when the communists took over Cuba, the danger became real. Until the late 1970s, Ellington aircraft—initially F-86s, but by 1961 supersonic Century-series F-102 and F-101 interceptors—were continuously patrolling the Gulf.

In 1959, the national headquarters of the Civil Air Patrol, the Air Force's civilian auxiliary, moved to Ellington. It was located in one of the World War II wood-frame headquarters buildings, quickly nicknamed "the Ponderosa" due to its resemblance to the ranch house in the then-popular television show *Bonanza*. In 1967, CAP Headquarters obtained an IBM computer to track personnel. Since the headquarters building lacked air-conditioning, IBM refused to put the computer there. CAP headquarters moved to Maxwell Air Force Base, Alabama, and an air-conditioned building.

The year 1962 saw another addition to Ellington when NASA came to town. The Manned Spaceflight Center was being built in the Clear Lake area, 10 miles south of Ellington. The space agency needed a place to operate its aircraft, and—temporarily, until the MSC was finished—for its laboratories and offices. The stay at Ellington was intended to last only two years. But like many "temporary" arrangements, NASA stayed at Ellington "for just two more years" until 1979. Then, finally accepting that it was there to stay, NASA refurbished or replaced the quarters it had occupied since the 1960s.

As Ellington entered the 1970s, the Air Force was shrinking. Ellington had been thrown up in 1940–1941. Most buildings lacked air-conditioning. Wood was rotting due to heat, humidity, and insects. The Air Force began transferring Ellington training schools to bases with newer, more modern buildings. The need for it as an air defense base on the Gulf Coast was declining. The Air Force wanted out of Ellington as early as 1966. Only NASA and the Texas Air National Guard forced the Air Force to keep the base open.

In 1976, the Air Force pulled the plug. They gave the base to the Texas Air National Guard (TANG), letting the state operate the base, with the TANG and NASA as the main tenants. For the next eight years, the TANG ran Ellington, once again Ellington Field.

Ellington's future was debated. NASA thought about staying two more years. The TANG liked Ellington, but it was too much base for their needs. Houston thought Ellington might be useful, but it was close to the then-underutilized Hobby Airport. No one wanted to take charge.

In 1981, the Reagan administration announced it was considering using barracks at Ellington still owned by the federal government to house Cuban and Haitian refugees. That plan fell through, but the scare convinced Houston to buy Ellington. Negotiations took three years, as Houston and Pasadena fought over which city got the facility. In a Solomonic decision, the federal General Service Administration divided the baby. Houston got the airfield, and Pasadena got Ellington's 107-acre golf course.

Ellington Airport became the Houston Airport System's (HAS) general aviation airport, but the city needed time to sort out what that meant. For a decade, things remained much the same as they had been when the city bought Ellington. NASA remained and upgraded its facilities. The TANG remained. Houston began encouraging Ellington's use for commuter flights to Intercontinental Airport on Houston's north side and used it as a contingency field for both Hobby and Intercontinental against weather emergencies.

In the 1990s, HAS began upgrading Ellington. Remaining World War II buildings were torn down. New general aviation hangars and facilities went up. New tenants were found, notably the Lone Star Flight Museum, which opened in 2017. A new control tower went in. Finally, in 2006, Ellington became the proposed site of Houston Spaceport, with spacecraft using Ellington's runways to reach orbit. Construction of a spaceport facility on Ellington's undeveloped south side began in 2019.

The importance of air defense was highlighted in 2001 after the September 11 attacks. Between 1976 and 2001, the 111th operated F-4 and F-16 fighters out of Ellington. Suddenly, those F-16s looked really attractive. While yet another round of base closings threatened Ellington, it ended in 2008 with Ellington being tapped to be the site of a multi-service reserve base. Instead of shrinking, its military presence began expanding in 2009.

Ellington's future in 2020 looks bright—as bright as it did in 1917 when it first opened.

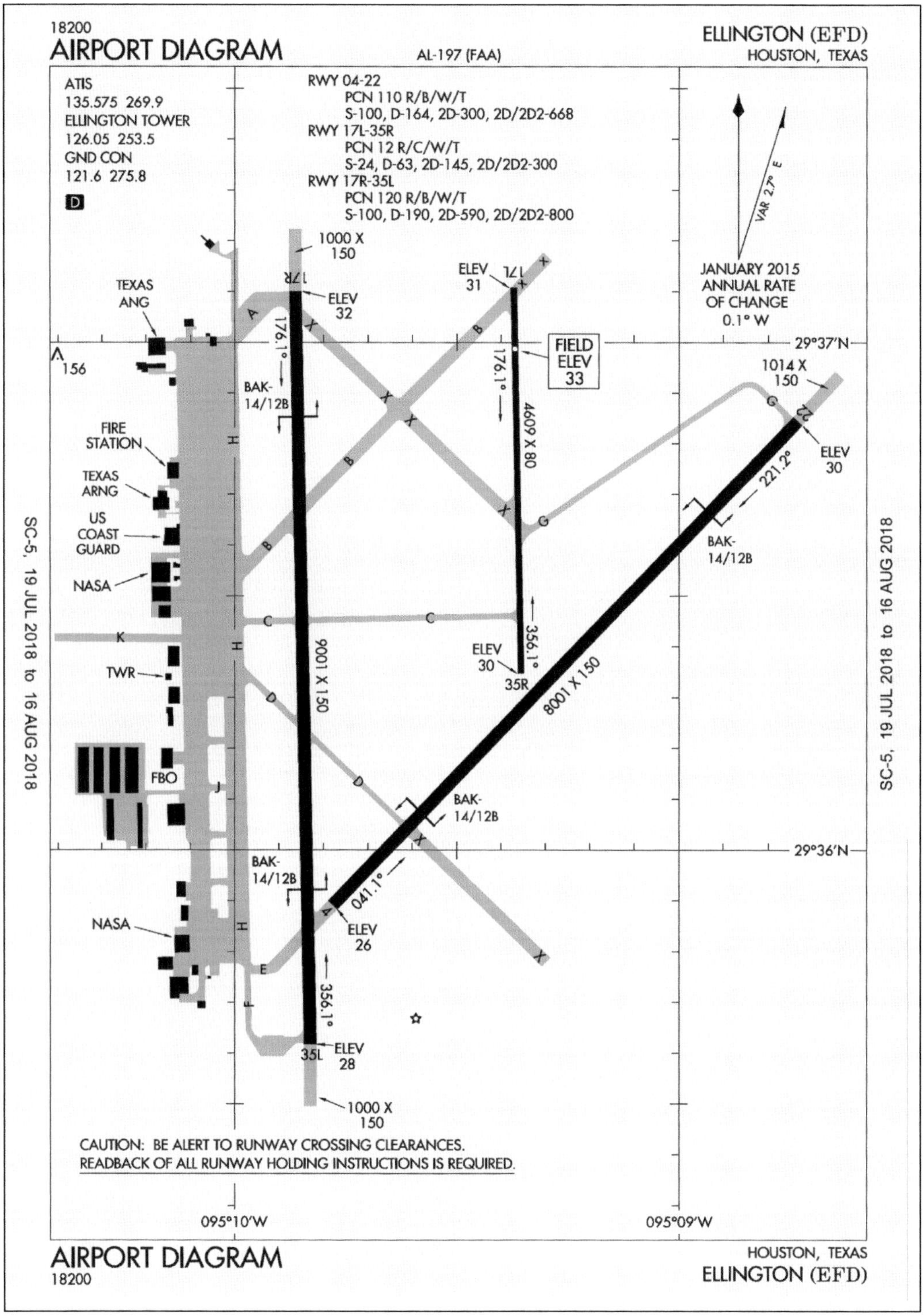

This FAA airport diagram for Ellington Field was for pilots using the airport. Black indicates runways or buildings, gray indicates taxiways. The main runway, W5L/17R, is 9,000 feet long and 150 feet wide. (FAA.)

One

Origins and World War I

1917–1927

On April 6, 1917, the United States declared war on the German Empire, entering World War I as an associate power of the Allied Powers. This new thing, the airplane, was playing a big part in the war.

In December 1916, the US Army had 12 qualified pilots. Only months later, the French asked the United States to contribute 4,500 aircraft, 5,000 pilots, and 50,000 mechanics to the war effort by the spring of 1918. The US Army suddenly needed a *lot* more pilots. That meant a lot of new training airfields.

The Texas Gulf Coast was ideal for airfields. Flat, largely treeless, and mostly undeveloped, it had mild weather most of the year. That meant more hours in the air on an airfield that could be built quickly and cheaply. The Army purchased 1,280 acres of cow pasture southeast of Houston for one such field. It was named Ellington Field after Eric Ellington.

The next two years saw explosive growth at Ellington. Pilots and air gunners trained there. Schools for aircraft mechanics and radio operators sprang up. Classes in every aspect of flight were taught at the Advanced Flight, Bombing, and Navigation Schools established at Ellington. This included the first course in formation flying offered in the United States.

Nor did Ellington focus exclusively on the flying part of aerial warfare; it also experimented with air defense technology, testing sonic tracking for early warning of incoming aircraft. It failed to pay off, and another 20 years would pass before radar solved that problem.

Ellington saw night flying and air ambulances developed. A camp newspaper, *Tale Spins*, emerged. After World War I, in July 1919, there was even a mid-air wedding at Ellington when a Lieutenant Meade stationed there took advantage of the arrival of a four-engine Handley-Page bomber to wed a Miss Dumont 2,000 feet over the airfield.

This chapter tells these stories, and follows the development of Ellington from a cow pasture to one of the nation's largest airfields.

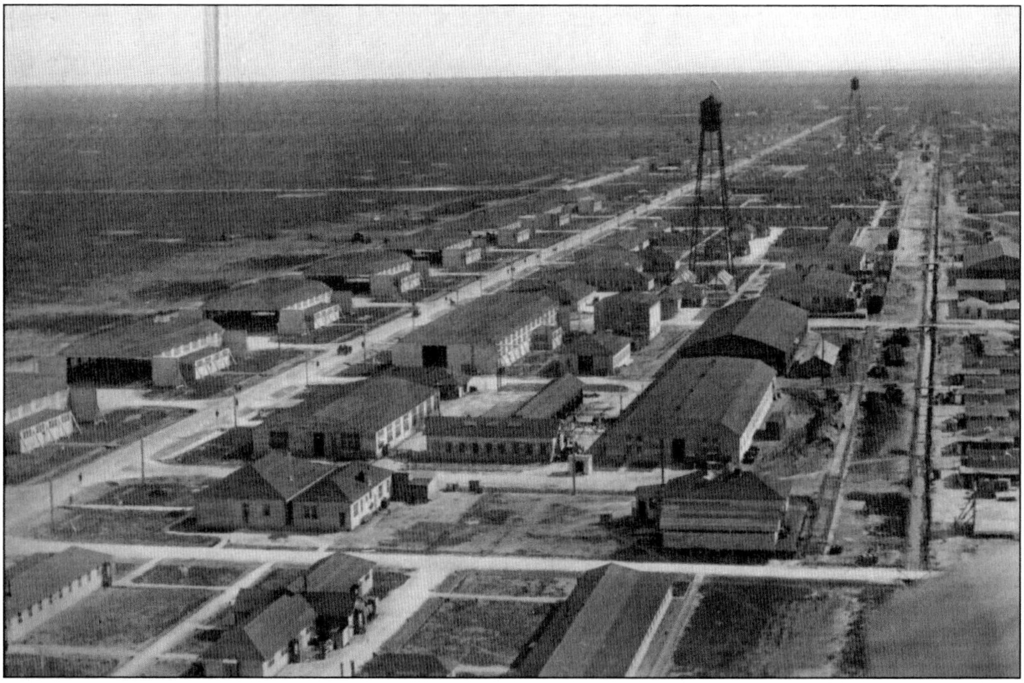

Ellington Field is seen here in late 1918. Three years earlier, it had been a cow pasture. By the end of World War I, it was filled with rows of hangars, workshops, offices, and barracks, although many soldiers still slept in tents. (LSFM.)

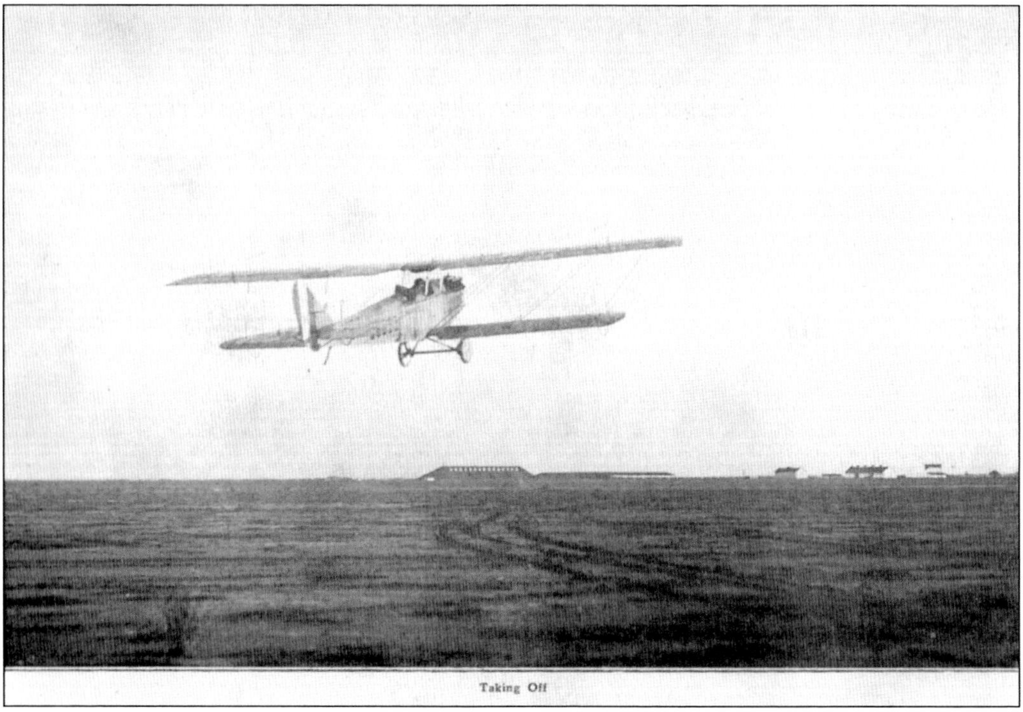

A training aircraft takes flight from Ellington. The runways were largely grass. Some of the airfield's hangars can be seen in the background. (LSFM.)

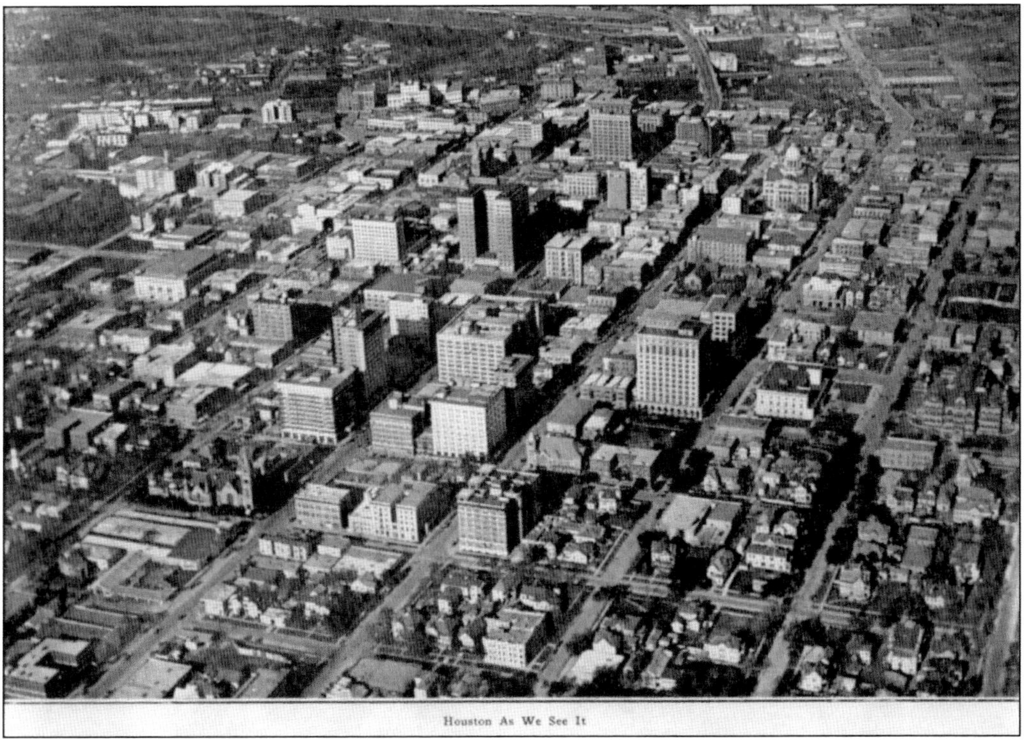

Houston in 1918 was a city of just over 100,000 inhabitants. This is a picture of downtown Houston (and a few then-outlying suburbs) taken from an aircraft stationed at Ellington in 1918. (LSFM.)

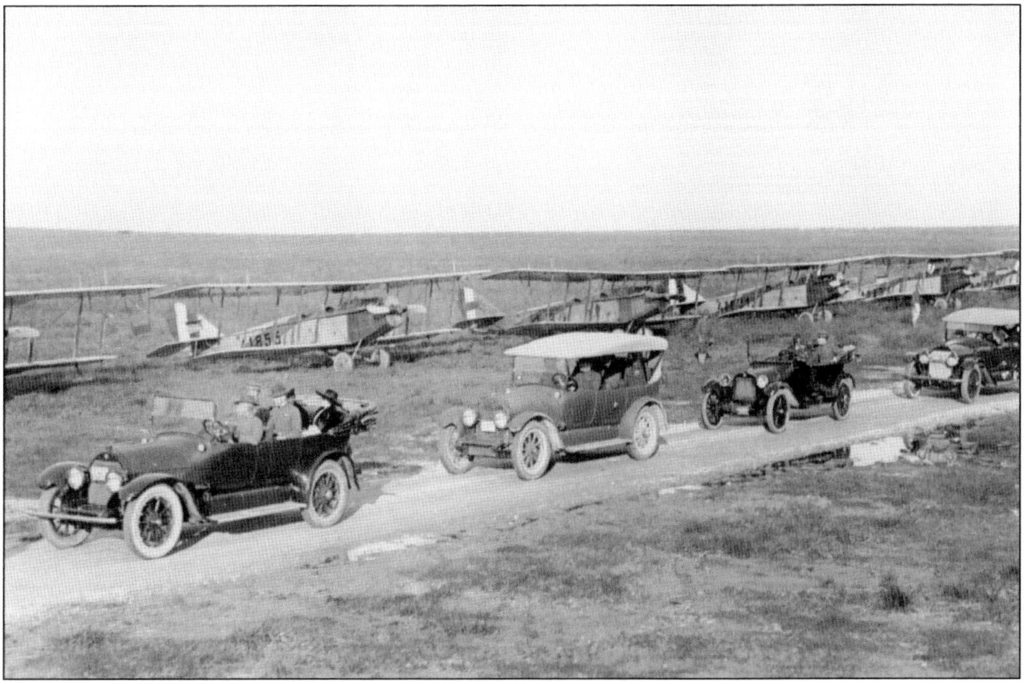

A parade of automobiles occupied by officers, notable civilians, and wives drives along an access road to inspect a line of Curtis "Jenny" training aircraft. (DVIDS.)

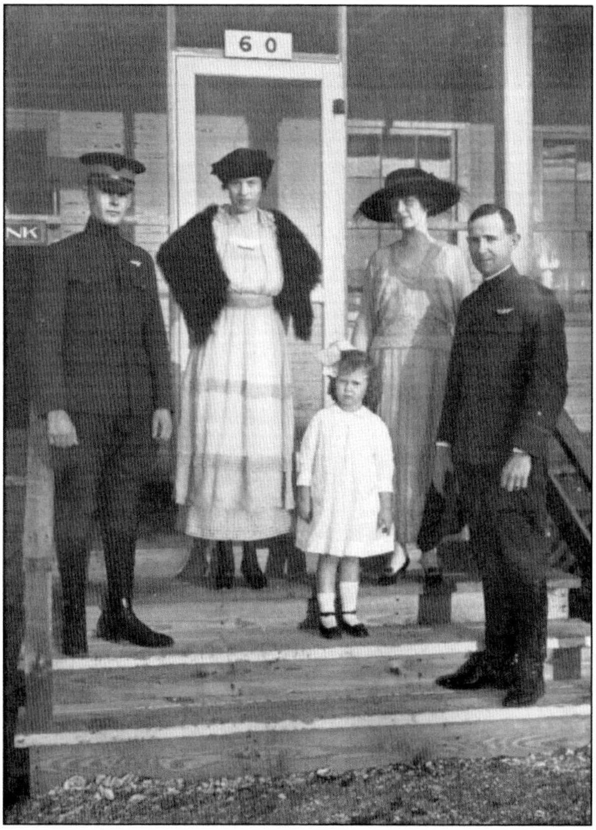

Ellington Field was commanded by Col. Ira Rader during World War I. He is shown here, at left, in front of his quarters. Next to him is his wife. On the right is Rader's deputy, Maj. Walter Frank, with his wife and daughter Hazel. (LSFM.)

The headquarters staff at Ellington during World War I poses in front of the headquarters building. (LSFM.)

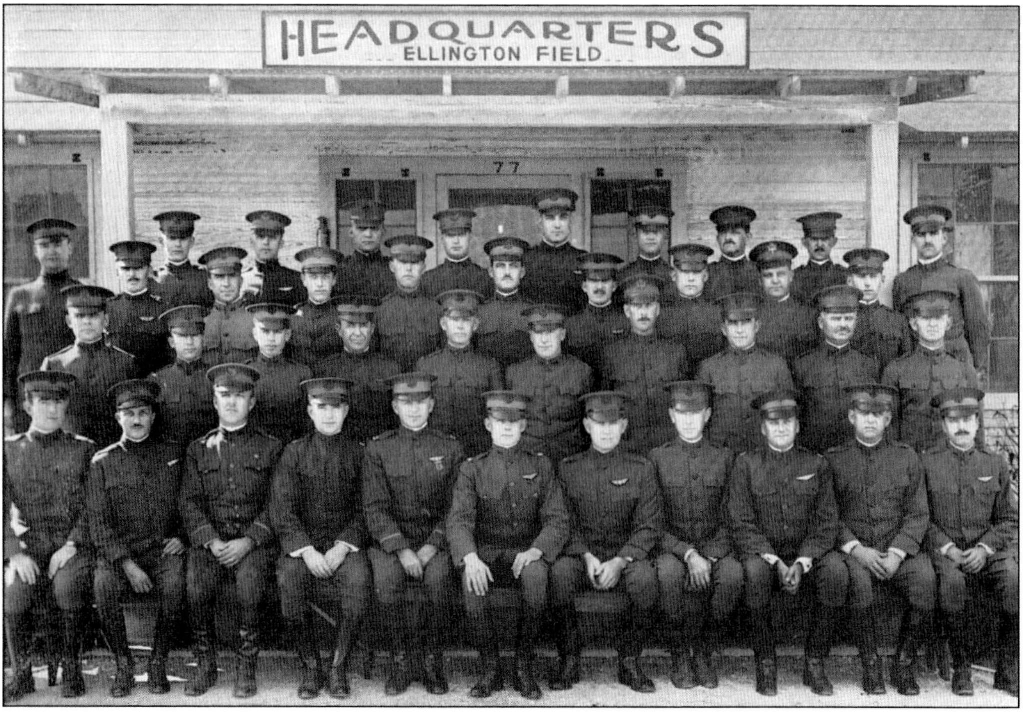

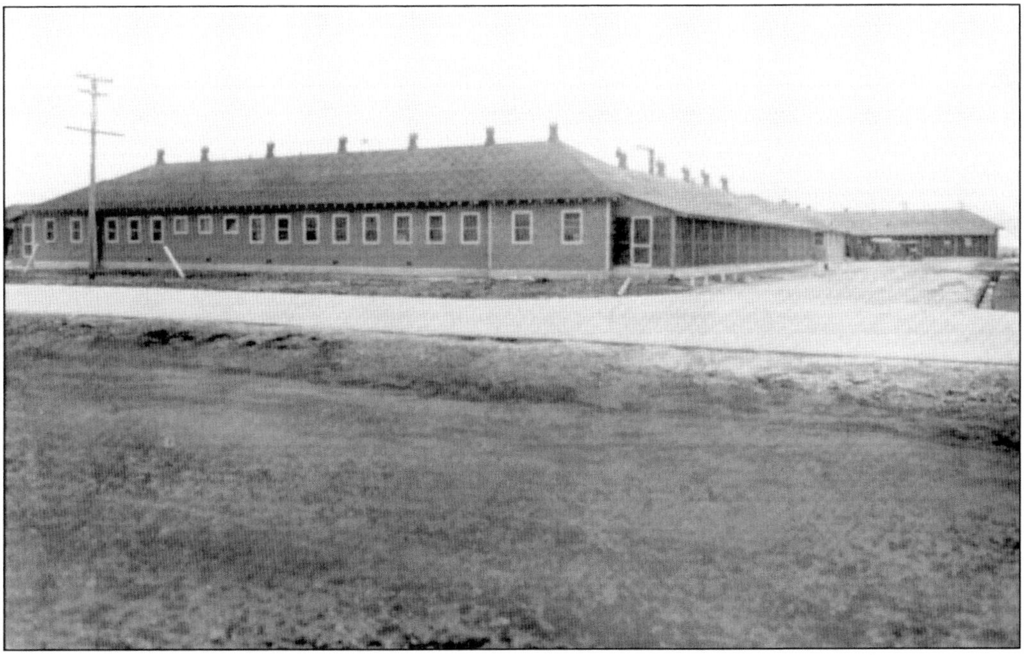

The post hospital is pictured in 1918. As was common practice during that period, it was a one-story wooden building with large windows and a screened-in porch to allow good ventilation in the days before air-conditioning. (LSFM.)

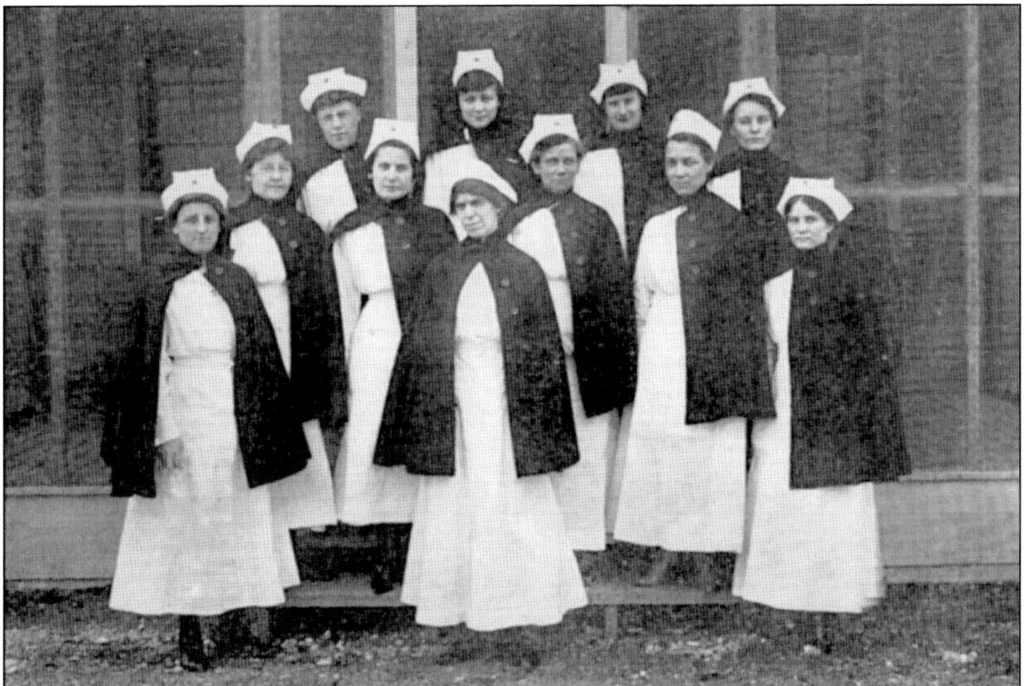

The hospital's nursing staff poses in front of the entrance to the porch. They appear experienced and reserved, wearing nursing uniforms that place an emphasis on projecting competence and formality. (LSFM.)

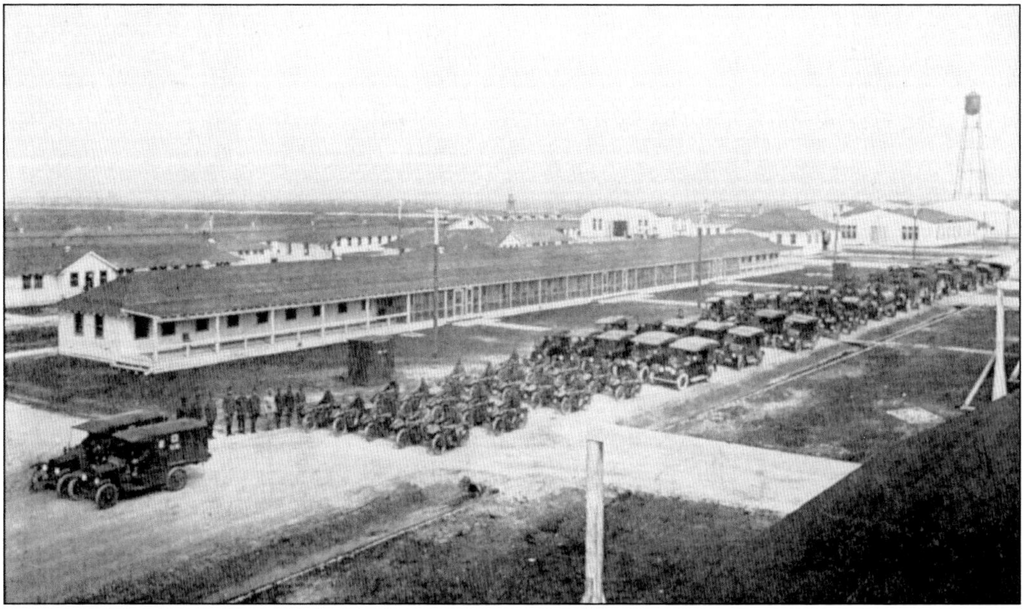

The Army Air Service was much more mechanized than the rest of the US Army of that time. Here, the start of a motorized road march from Ellington is pictured. (LSFM.)

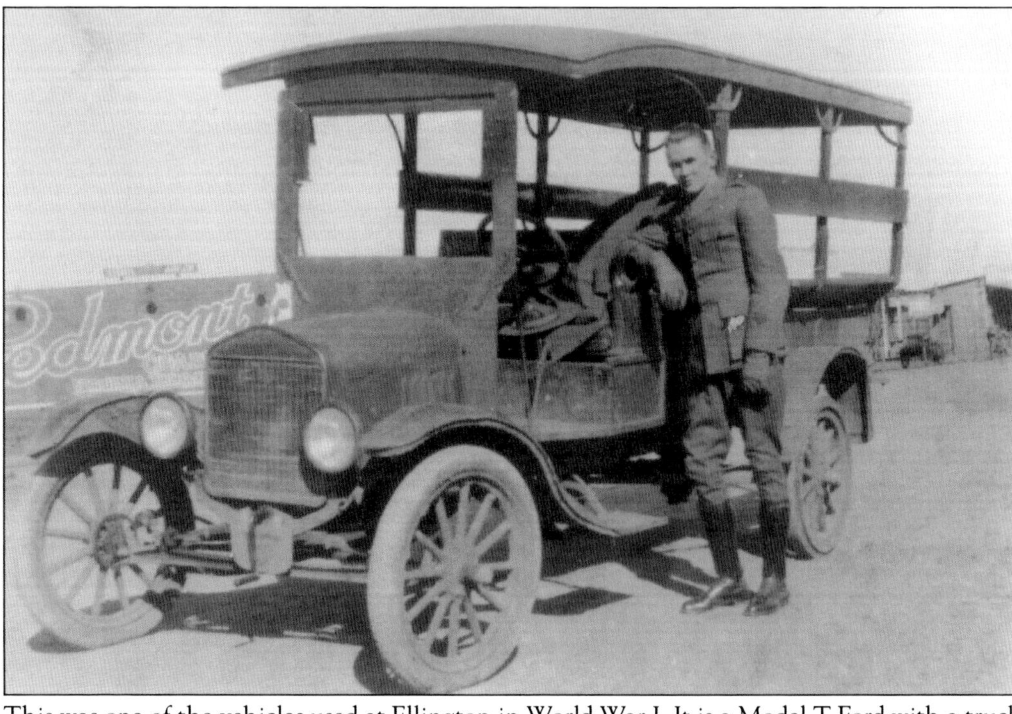

This was one of the vehicles used at Ellington in World War I. It is a Model T Ford with a truck body added. It has a framed back end, with benches running the length of the bed. It could be used to haul spare parts and tools or serve as a bus. (147th.)

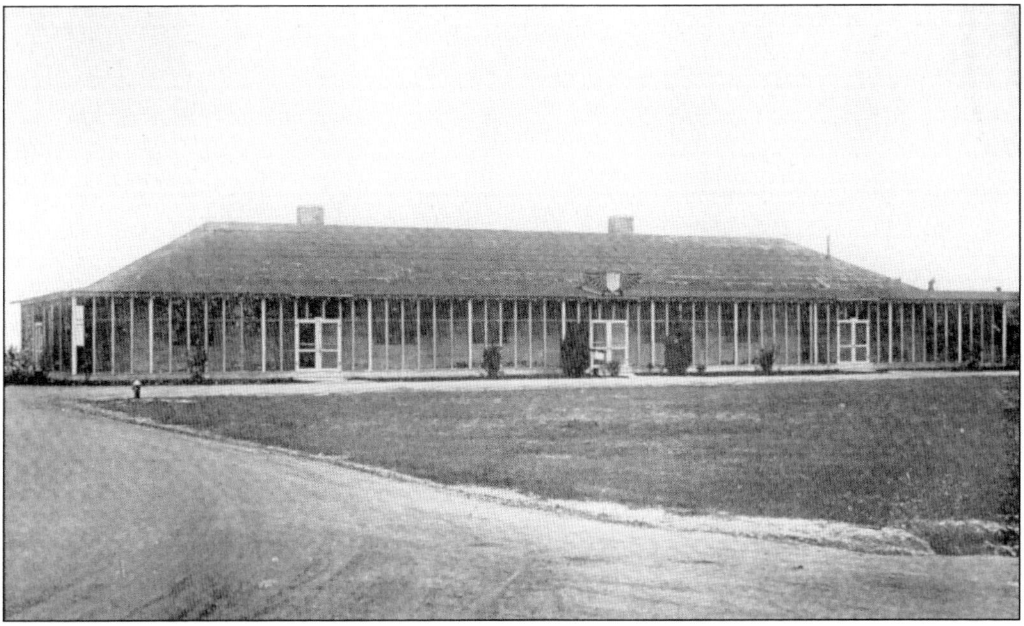

An air service flies on its stomach as much as an army marches on its stomach. This was the enlisted men's mess hall at Ellington. It is basic: bench seating at long plain tables in spartan settings. (LSFM.)

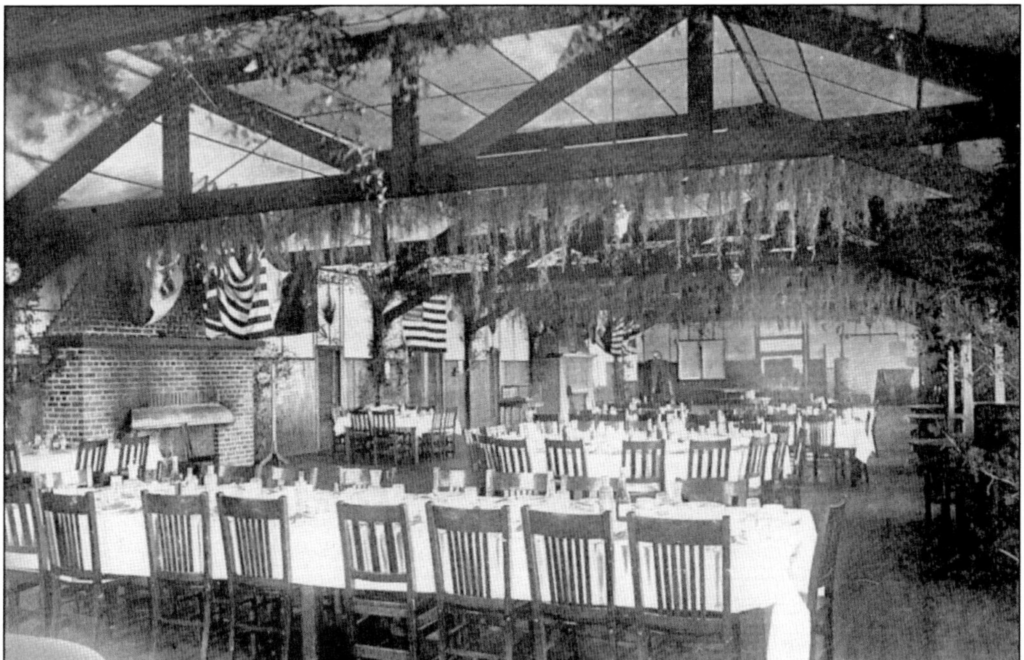

This is Ellington's officers' club, featuring tablecloths, individual chairs, and decorative plants. Although the officers dined with more luxury, their meals were eaten more formally than at the enlisted mess. The officers had to behave as gentlemen. (LSFM.)

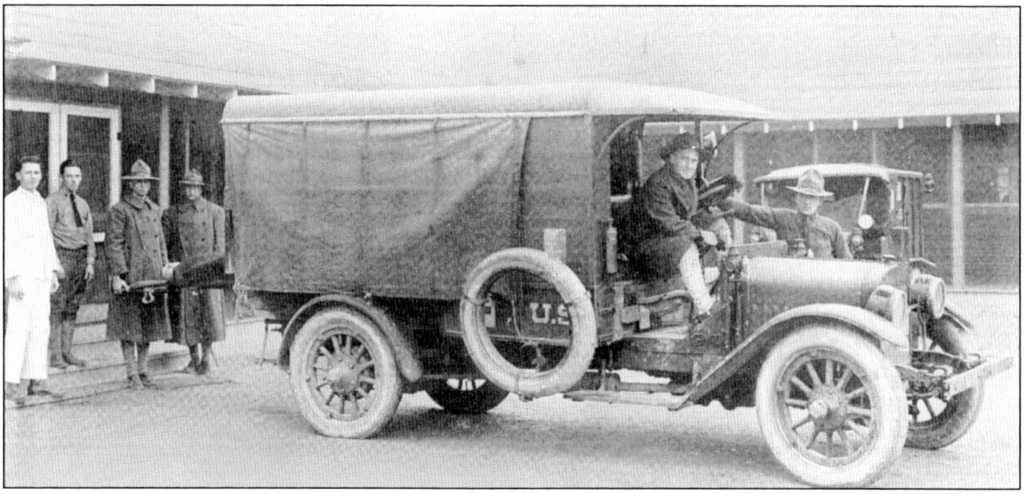

This ambulance was used to rush patients to the base hospital. (LSFM.)

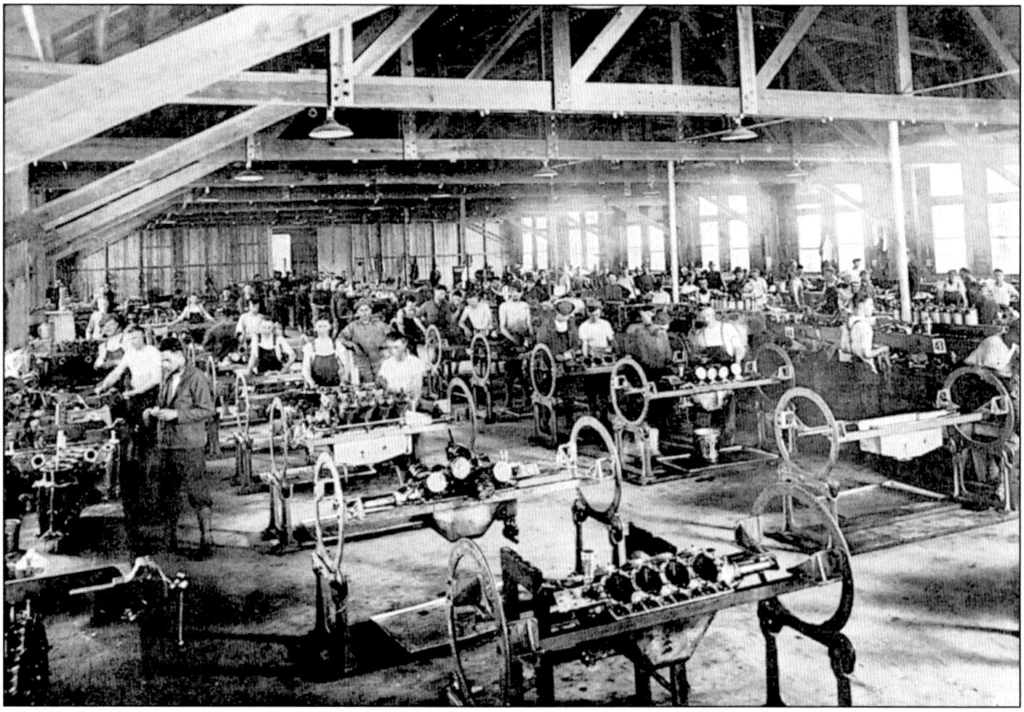

Ellington's purpose lay with the aircraft flown from there. At the time, the aircraft were shipped in crates and had to be assembled. Ellington had several buildings devoted to assembly and maintenance, including this one, the engine shop. These men appear to be assembling the OX-5 V-8 engines used on Curtis Jenny trainers. (LSFM.)

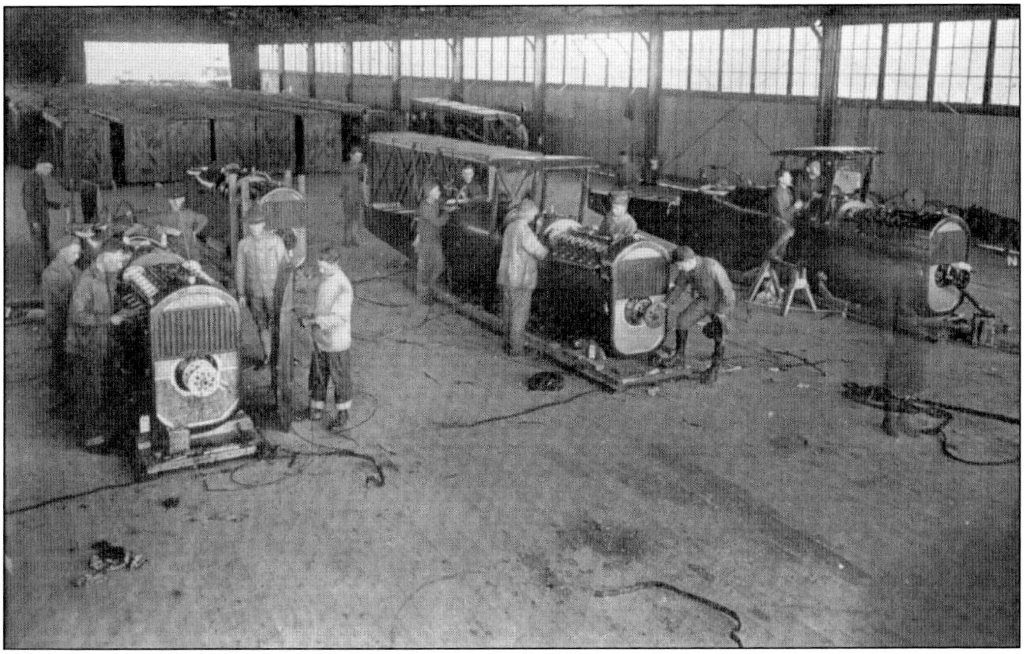

Another aircraft operating out of Ellington in 1918 was the de Havilland DH-4, a bomber and scout aircraft. While obsolescent on the battlefield, it was a lot hotter than the Jenny, and great for bombing training. Here, DH-4s are being uncrated and assembled in one of Ellington's workshops. (LSFM.)

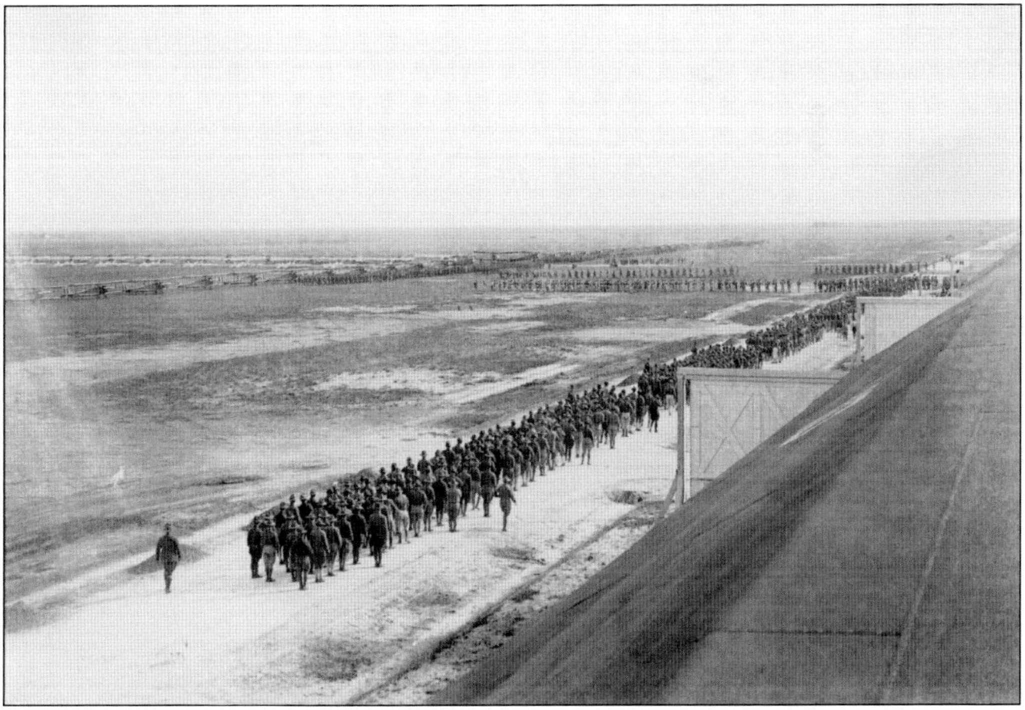

The ultimate purpose of Ellington was to train aircrew for the Army Air Service. Here, pilot cadets march out to their aircraft before a day of flight training. (DVIDS.)

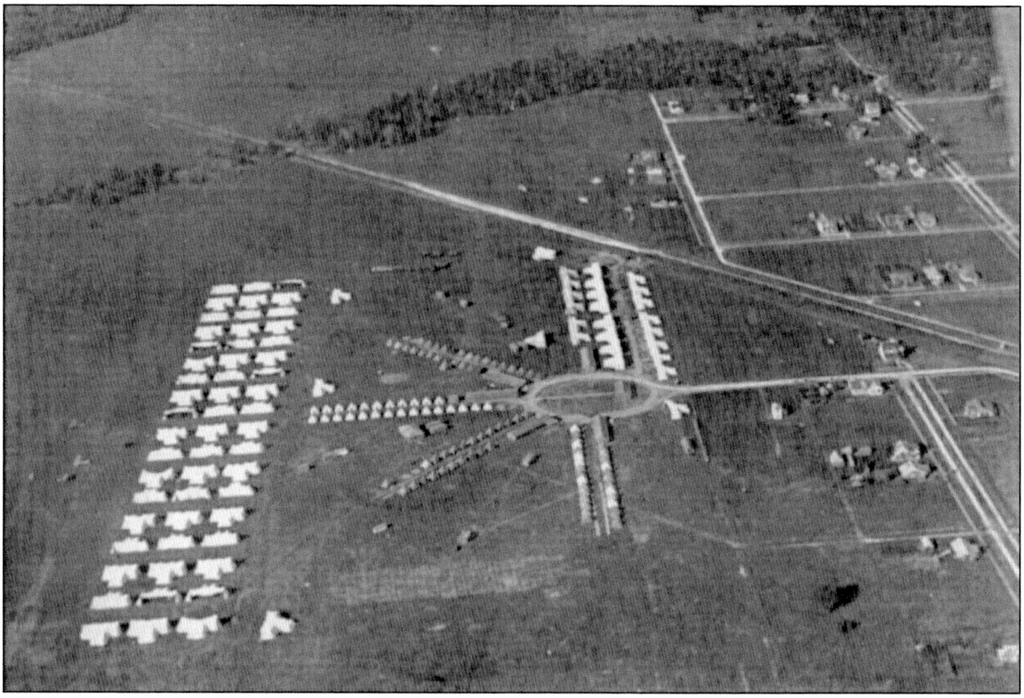

To disperse trainees and also to give student pilots somewhere to fly to, Ellington had several satellite airfields. One, shown here, was located at Park Place, 10 miles northwest of Ellington near where Loop 610 meets the Gulf Freeway. (LSFM.)

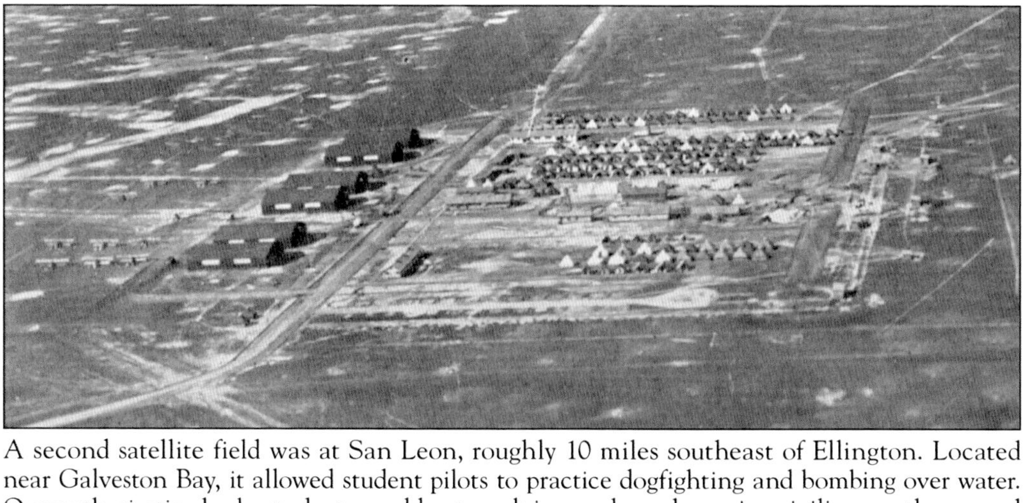

A second satellite field was at San Leon, roughly 10 miles southeast of Ellington. Located near Galveston Bay, it allowed student pilots to practice dogfighting and bombing over water. Overenthusiastic play by students would not result in wrecks endangering civilians on the ground if the aircraft were over the bay. (LSFM.)

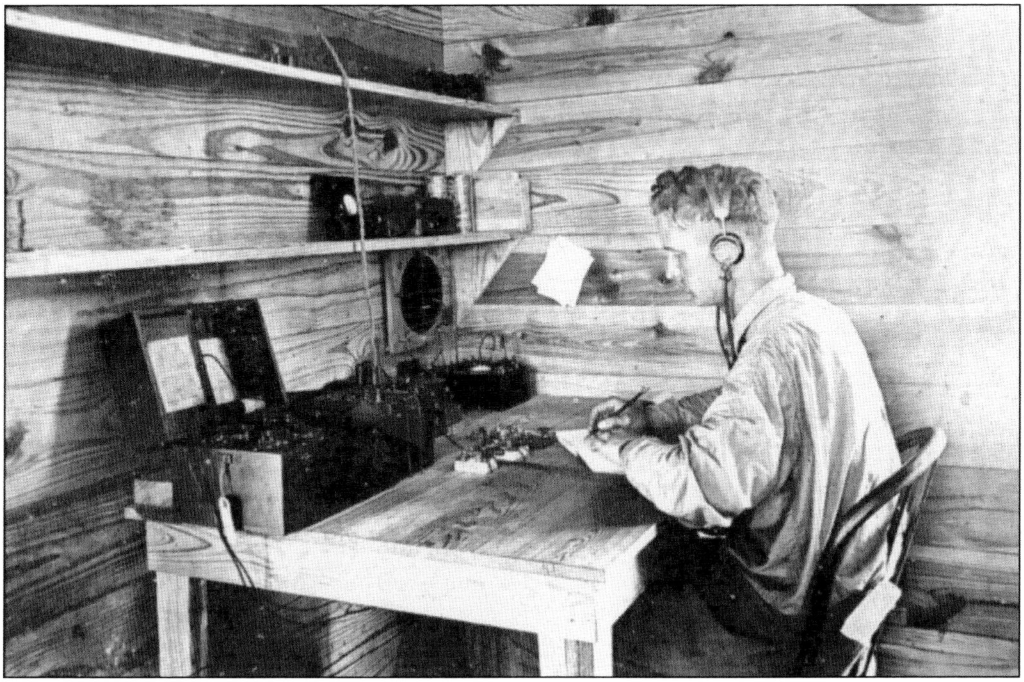

Flying was not the only thing taught at Ellington. Radio was then cutting-edge technology, and critical to aerial reconnaissance. This student practices with a radio set capable of flying in an aircraft. (LSFM.)

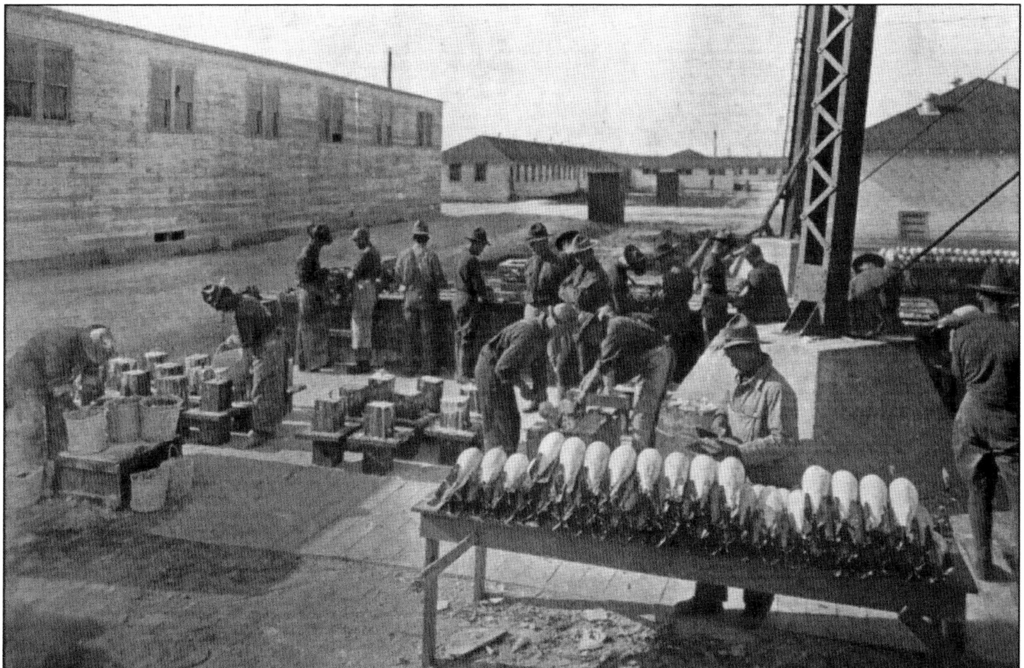

Bombing was also taught and practiced. Here, ground personnel fill practice bombs. They also paint the casings white, using whitewash to mark them as practice munitions. (LSFM.)

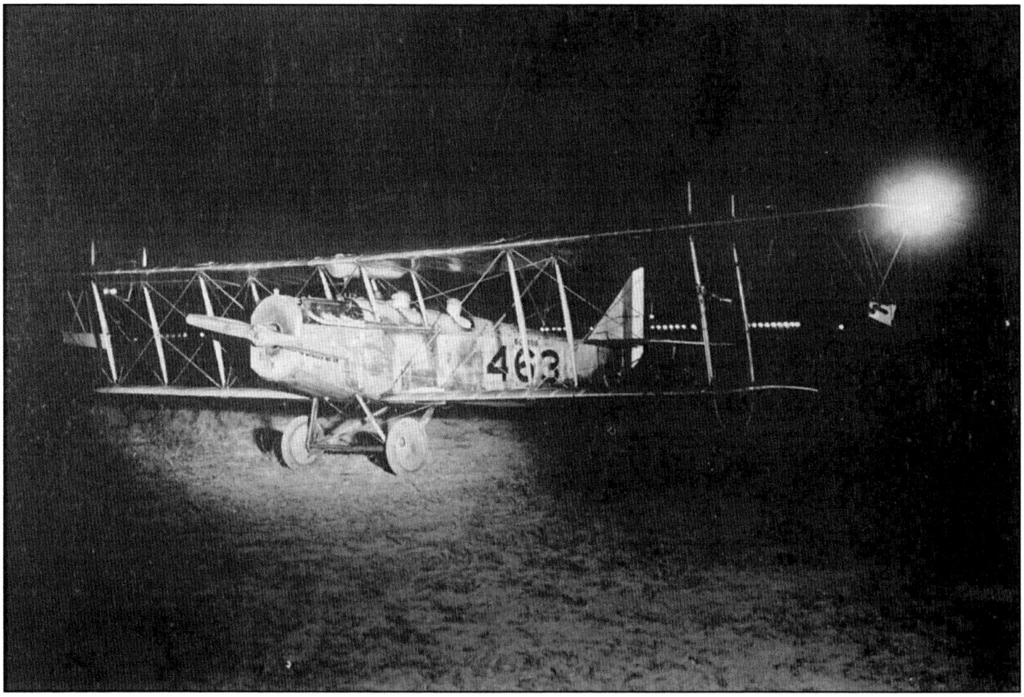

Ellington taught its students what were then cutting-edge aviation skills. This included night flying, a dark and mysterious art at the time. This Curtis Jenny is readying to take off from Ellington on a night flight. (LSFM.)

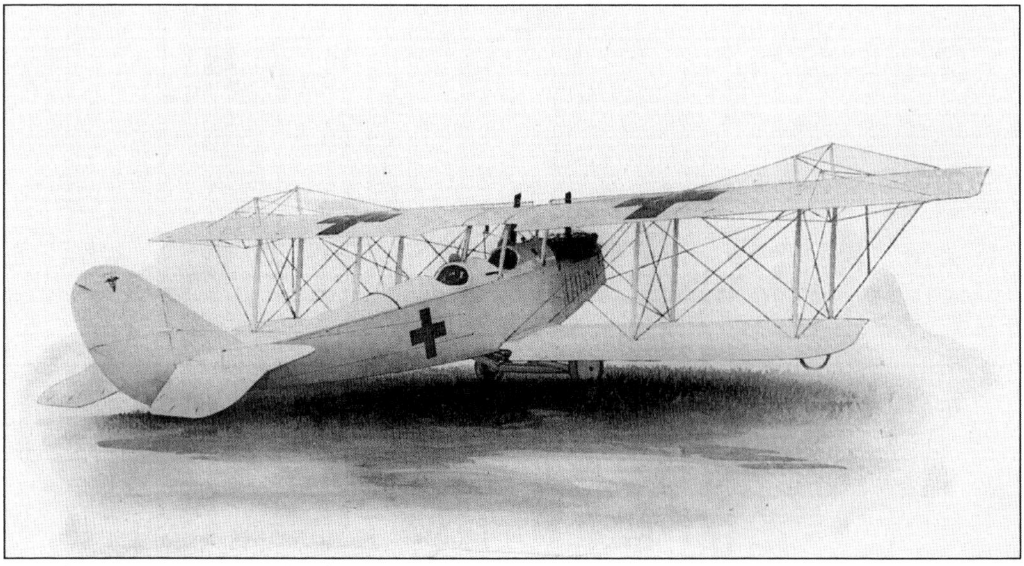

Another experiment involved air ambulances. This JN-4 was modified so the top of the rear cockpit could be removed and a stretcher patient carried. It was one of the earliest experiments in airborne casualty evacuation. (LSFM.)

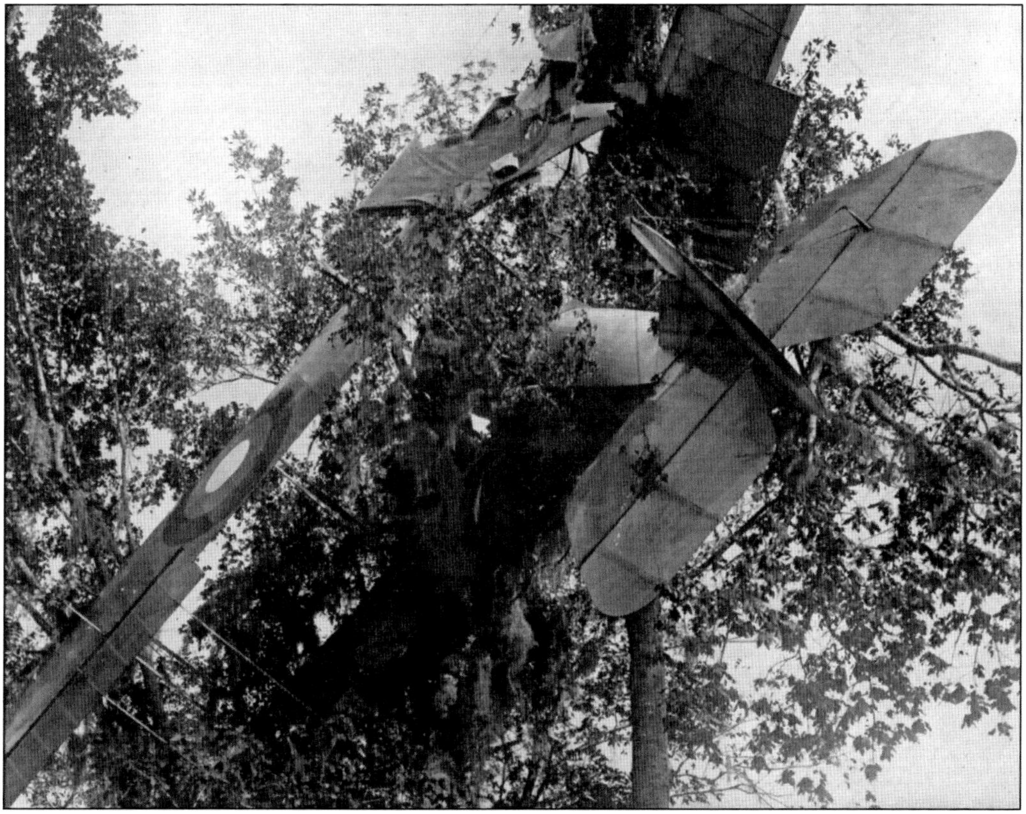

There was always at least one tree near every airfield that seemed to have strange magnetic power to attract aircraft. Or perhaps it was simply that too many student pilots lacked situational awareness when flying close to the ground. Whatever the reason, this Jenny stationed at Ellington ended up wrapped around this tree. (LSFM.)

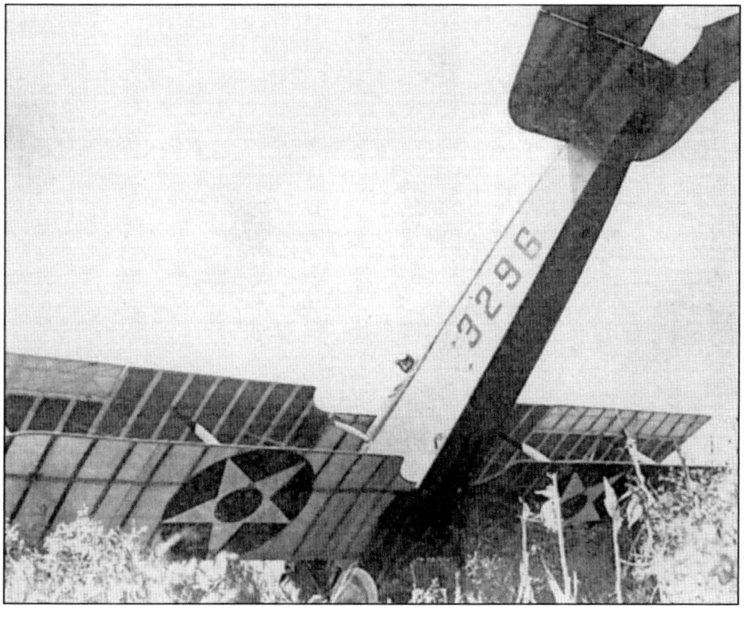

Or, as happened here, a student might misjudge the approach angle while landing, hit the brakes too hard, or just misjudge his motion, causing the Jenny to nose over. Fortunately, about the only way this type of accident could kill a pilot would be if they died of embarrassment. (LSFM.)

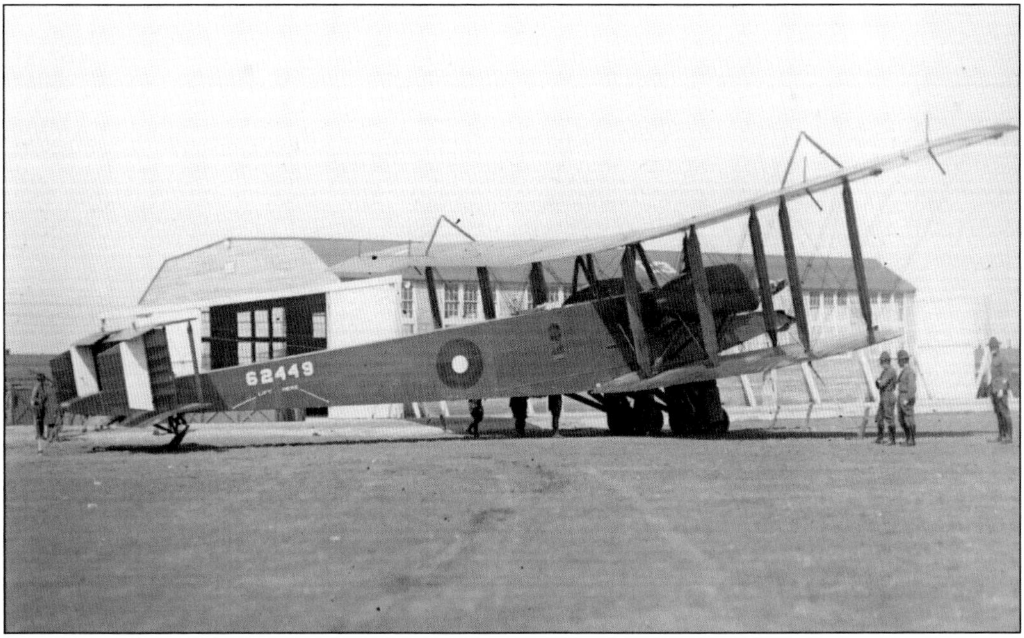

One of the more unusual aerial events at Ellington was an airborne wedding in 1919. When this massive (for the time) Handley-Page heavy bomber visited Ellington on June 20, 1919, one officer convinced his fiancée to tie the knot aboard the aircraft, in flight, several thousand feet above the ground. (LSFM.)

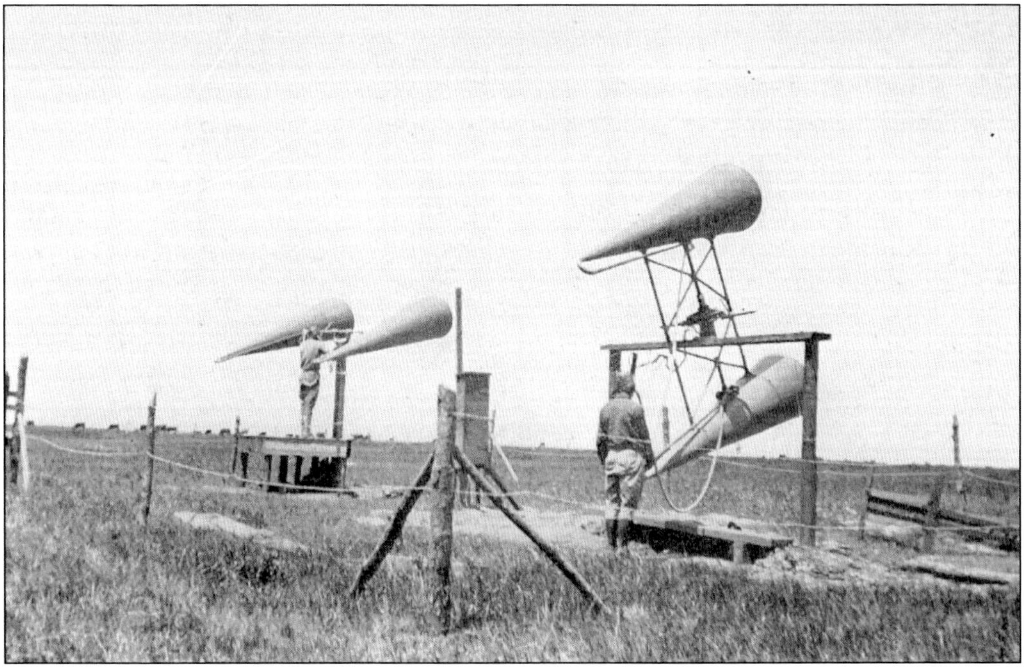

Radar did not exist until the 1930s, but the Army Air Service wanted ways to detect incoming aircraft as early as World War I. One technology investigated was sound detection. Ellington had several of these detectors to evaluate their reliability. One pair is shown here. The system sort of worked, although it was not very effective. (LSFM.)

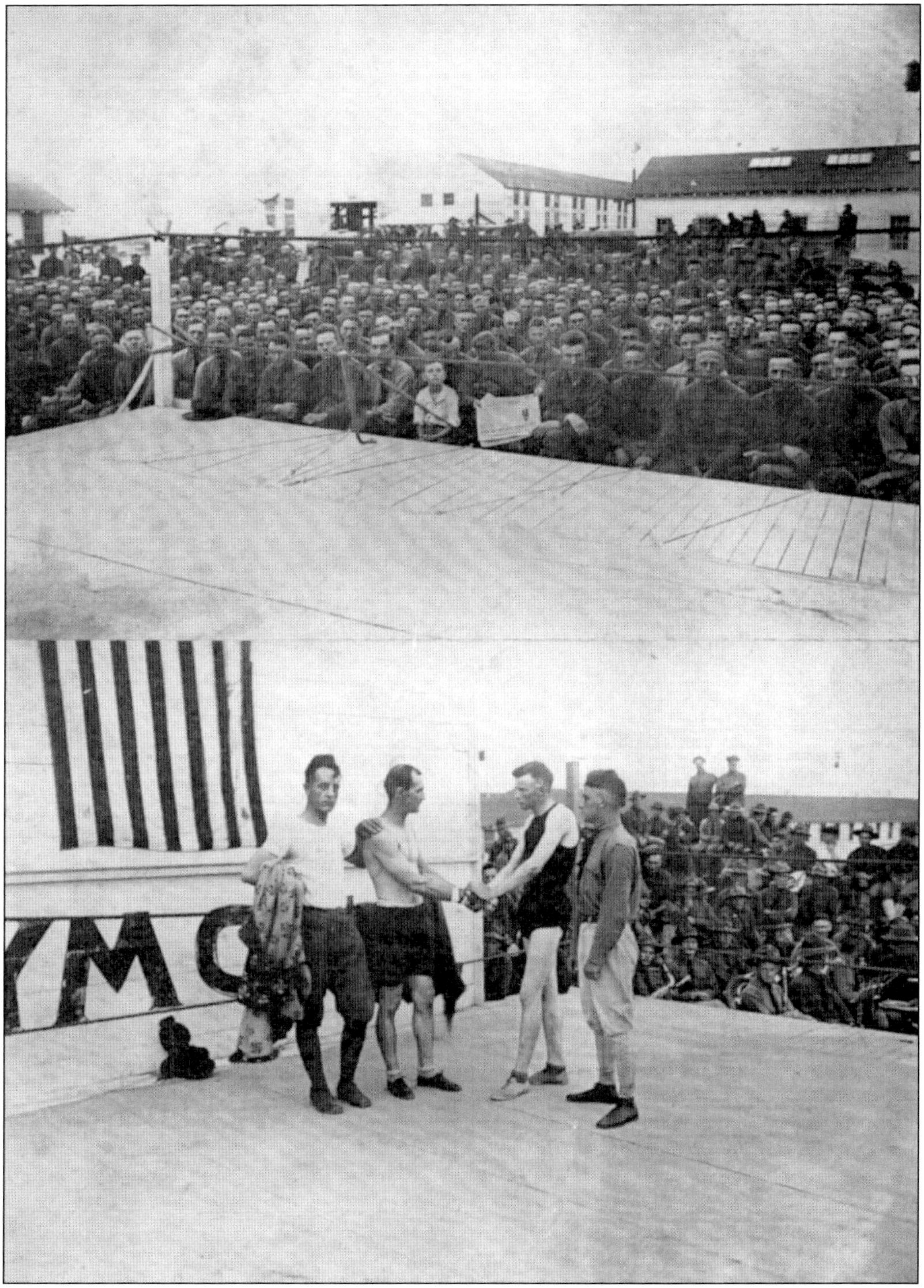

Sports were big at Ellington and throughout the Army during World War I. Boxing was especially favored, as officers believed it developed an offensive spirit. It was also popular among the troops, who turned out in numbers to watch a camp boxing match like the one pictured here. (LSFM.)

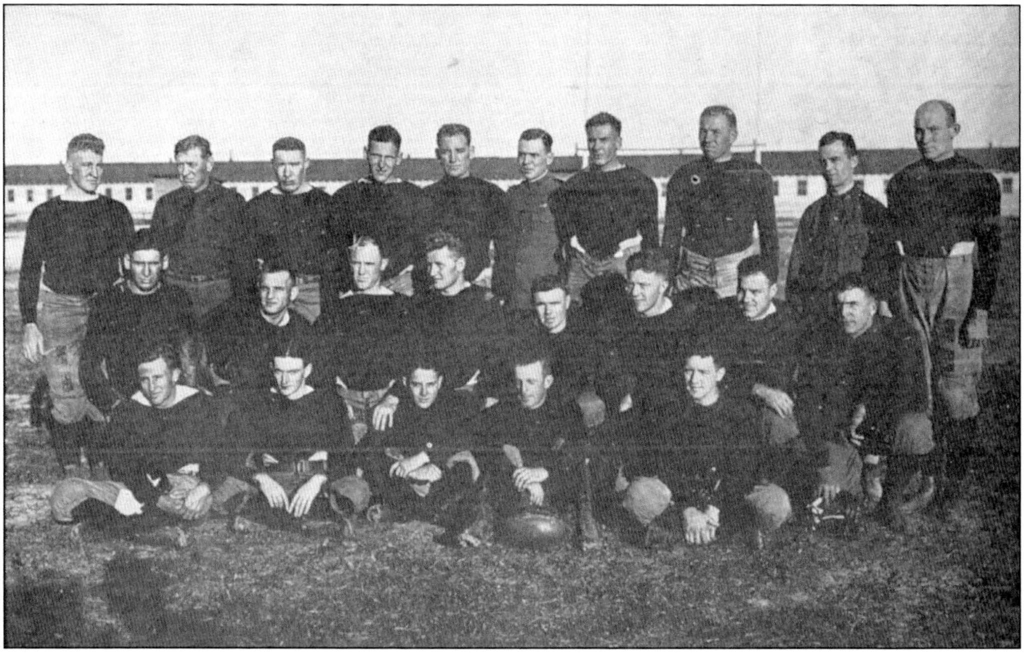

Football was another sport favored by the military. It was not only an aggressive sport, but it also promoted teamwork and cooperation, qualities as important on the battlefield as on the football field. This was the Ellington camp football team in 1918. (LSFM.)

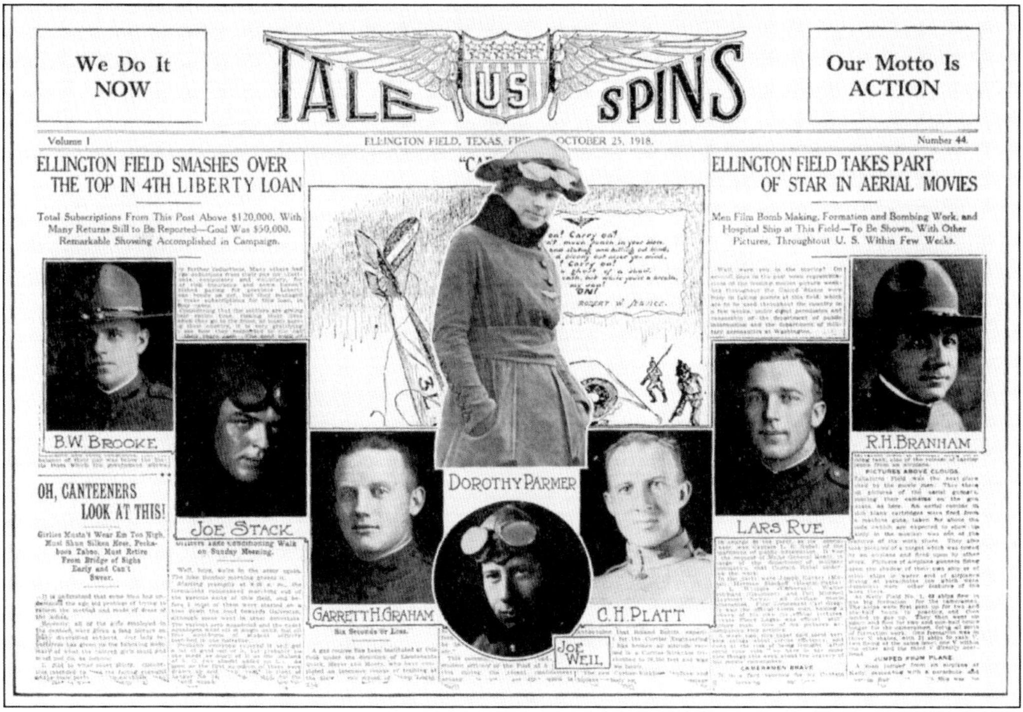

We Do It NOW

TALE US SPINS

Our Motto Is ACTION

Volume 1 — ELLINGTON FIELD, TEXAS, FRI… OCTOBER 25, 1918. — Number 44.

ELLINGTON FIELD SMASHES OVER THE TOP IN 4TH LIBERTY LOAN

Total Subscriptions From This Post Above $120,000, With Many Returns Still to Be Reported—Goal Was $50,000. Remarkable Showing Accomplished in Campaign.

ELLINGTON FIELD TAKES PART OF STAR IN AERIAL MOVIES

Men Film Bomb Making, Formation and Bombing Work, and Hospital Ship at This Field—To Be Shown, With Other Pictures, Throughout U. S. Within Few Weeks.

OH, CANTEENERS LOOK AT THIS!

Army bases, including Army airfields, were communities. In many ways they were like small civilian towns, and had many amenities found in small-town America, including a newspaper. *Tale Spins* was the Ellington Field newspaper in World War I. (LSFM.)

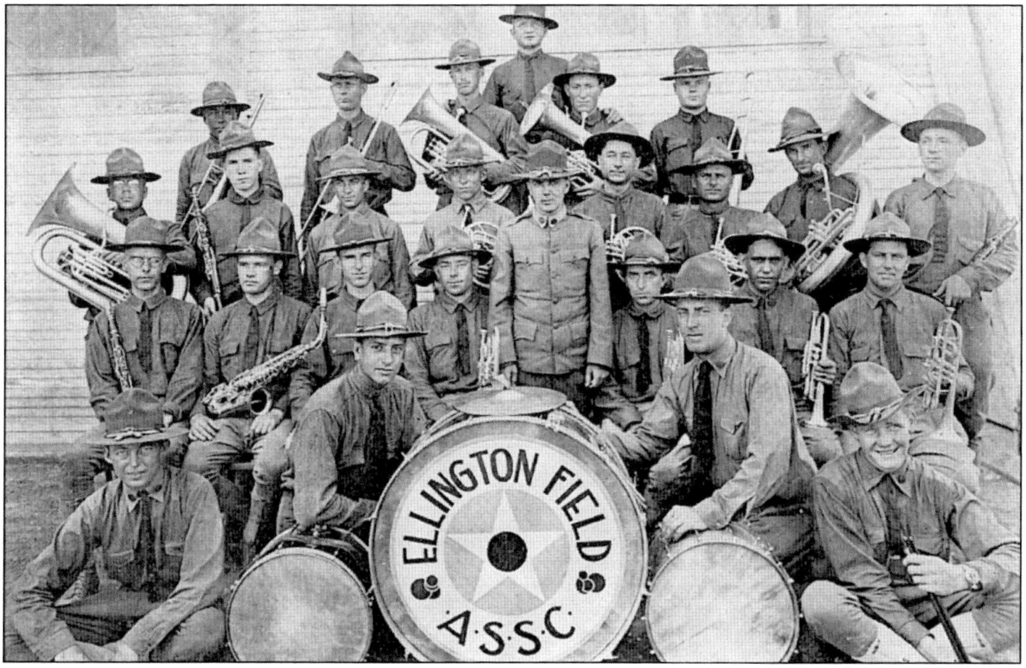

Music was also important to the military. It was still used to transmit orders in World War I, especially in camp, where "Reveille" started the day and "Taps" ended it. A band was also thought to build morale. This was the Ellington Field band in World War I. (LSFM.)

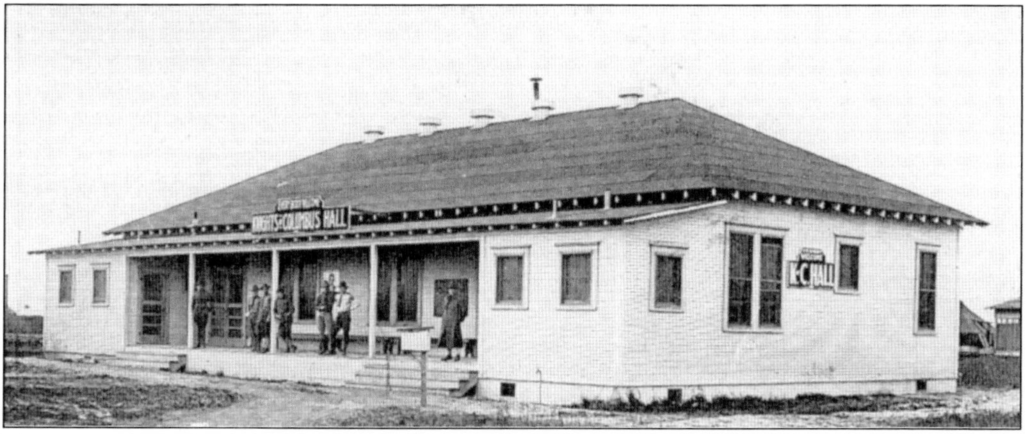

There were also places for troops to relax in their off time. Ellington had a Salvation Army, and several buildings housing fraternal organizations, including this Knights of Columbus hall. (LSFM.)

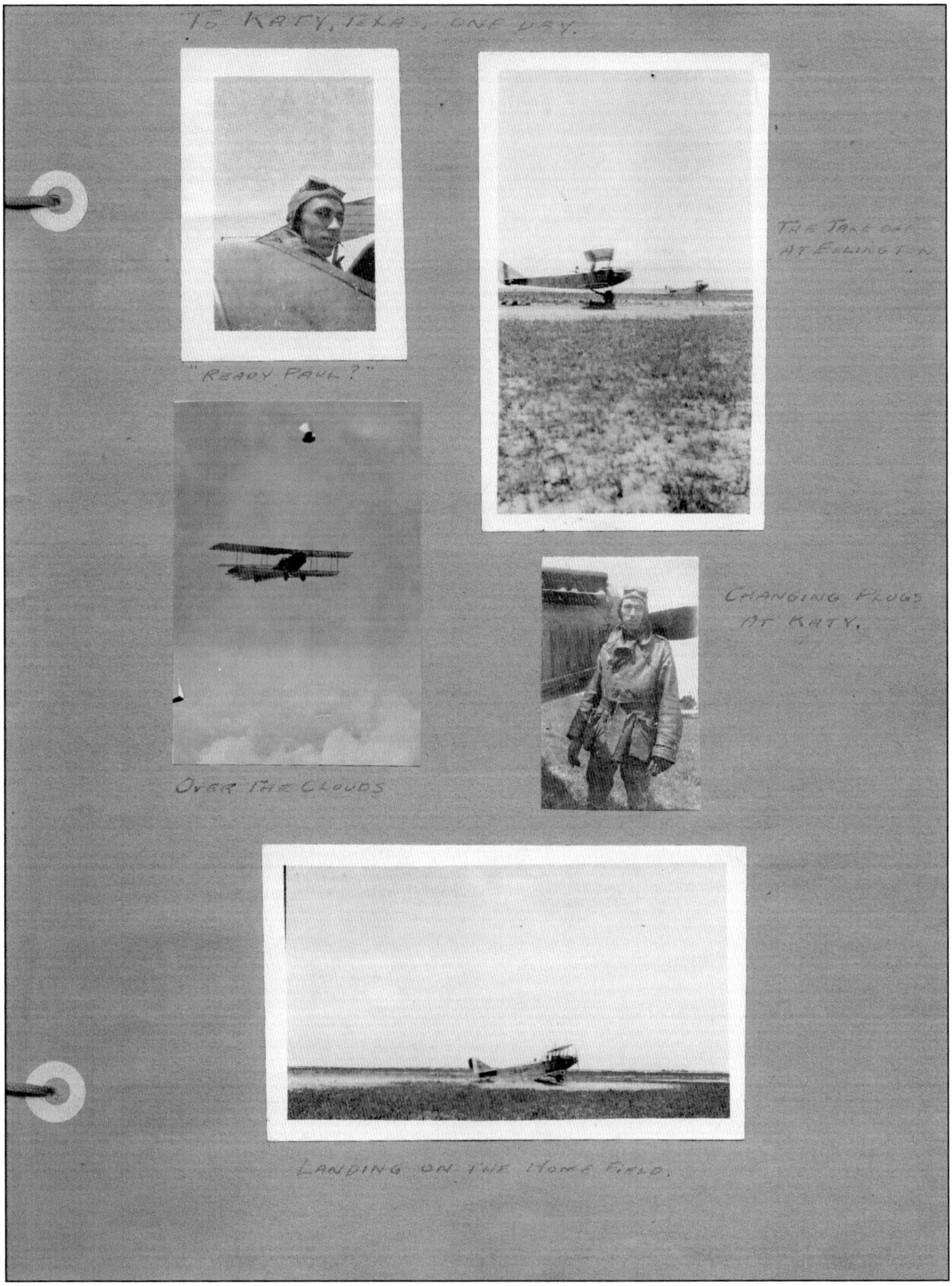

Andrew J. LaBoiteaux was a pilot who kept a scrapbook of his World War I Air Service experiences. After his death, the scrapbook went to the National Museum of the Air Force in Dayton, Ohio, where it is on display. While at Ellington, LaBoiteaux made a long-distance cross-country flight to Katy, Texas, about 50 miles away. This page includes his record of the trip. (NMUSAF.)

Two

World War II

1940–1946

By 1930, Ellington Field had reverted to nature. While the Army still owned the land, the aircraft had been sold off, the buildings razed, and rails and piping removed. Cattle again grazed on the former airfield. Only the runway marked it as ever having been an airfield.

Houston interests advocated reopening it as a military field to protect the area's oil industry. Instead, the Army based the local Air Guard units at Houston Municipal Airport (today's Hobby). World War II forced Ellington's reopening.

On November 23, 1940, Ellington was designated an Army Air Corps station. A week later, on December 1, 1940, it was officially reopened. Later that month, construction began transforming cow pastures back into an airfield. Drainage, sewers, and water supply were installed first. Wood frame buildings were erected next: rows of two-story barracks, five mess hall buildings, warehouses, office buildings, and a hospital. In all, 165 buildings went up. All were completed by April 1, 1941. This was followed by telecommunications, including 140 miles of telephone line.

Next, facilities for the aircraft went in. Paved concrete runways were laid, including a massive apron area in front of two large steel-reinforced hangars. The 42,000 square feet of sheltered space they provided was only enough to house aircraft being repaired or serviced, however.

Construction was complete by June 26, 1941. The first aircraft landed at Ellington on that date, and Ellington Field was officially declared operational. The 111th Observation Squadron, a unit stationed at Ellington during World War I and long associated with Houston and Texas, briefly stayed at Ellington during 1941 before being sent overseas and into combat. The first set of 50 pilot cadets reported for training on October 6, 1941. They got their wings on December 12, five days after Pearl Harbor brought the United States into the war.

They were the first of thousands of pilots, navigators, bombardiers, and preflight cadets Ellington would produce before World War II ended. By the end of 1943, over 15,000 students graduated from Ellington. This included a training squadron for black pilots. While segregated, it offered African Americans an opportunity to earn wings.

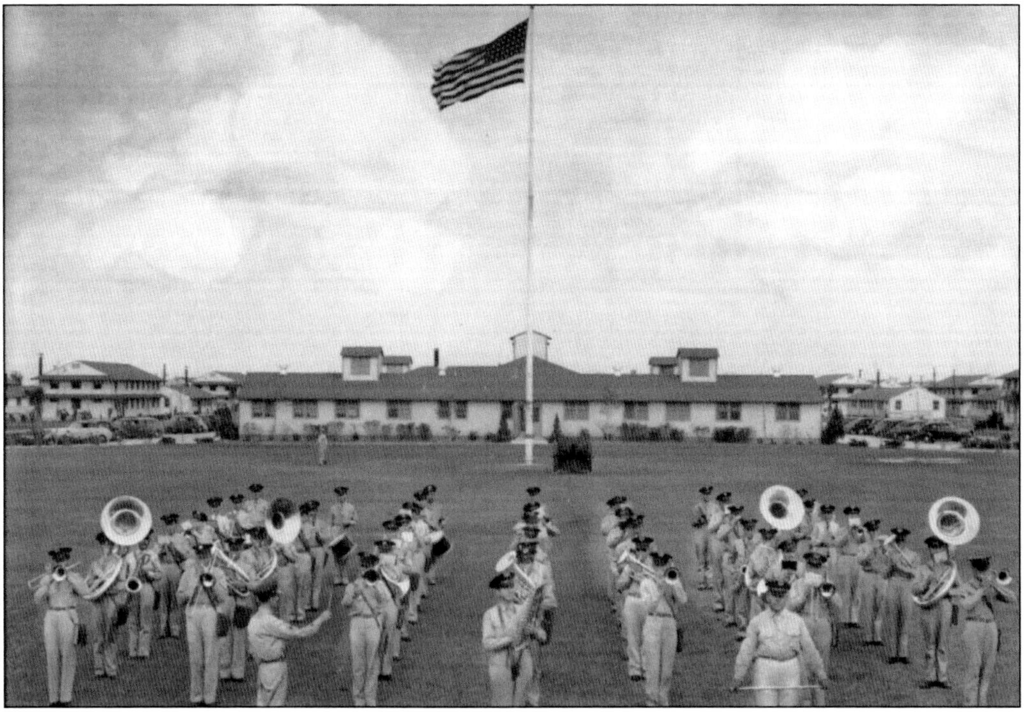

By 1940, Ellington had reverted to cow pasture. A year later, by July 1941, it looked like this: with 165 new buildings and a manicured parade ground. A massive construction effort had converted it into a state-of-the-art training airfield, bigger and better than it had been 22 years earlier. (LSFM.)

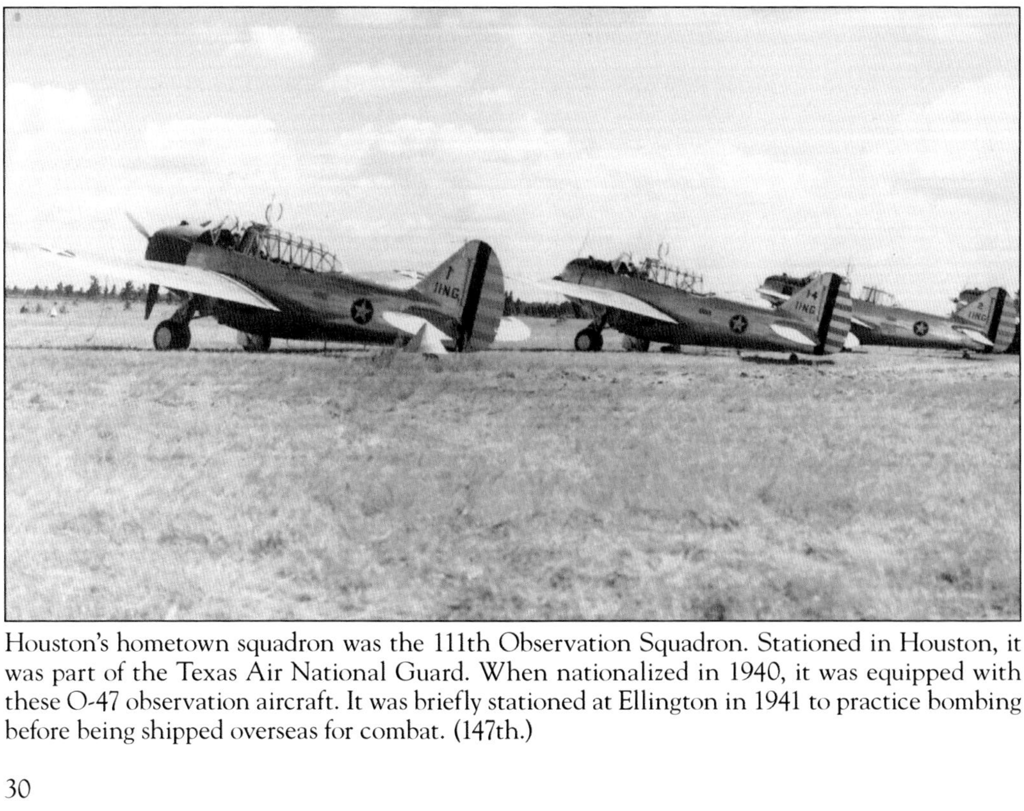

Houston's hometown squadron was the 111th Observation Squadron. Stationed in Houston, it was part of the Texas Air National Guard. When nationalized in 1940, it was equipped with these O-47 observation aircraft. It was briefly stationed at Ellington in 1941 to practice bombing before being shipped overseas for combat. (147th.)

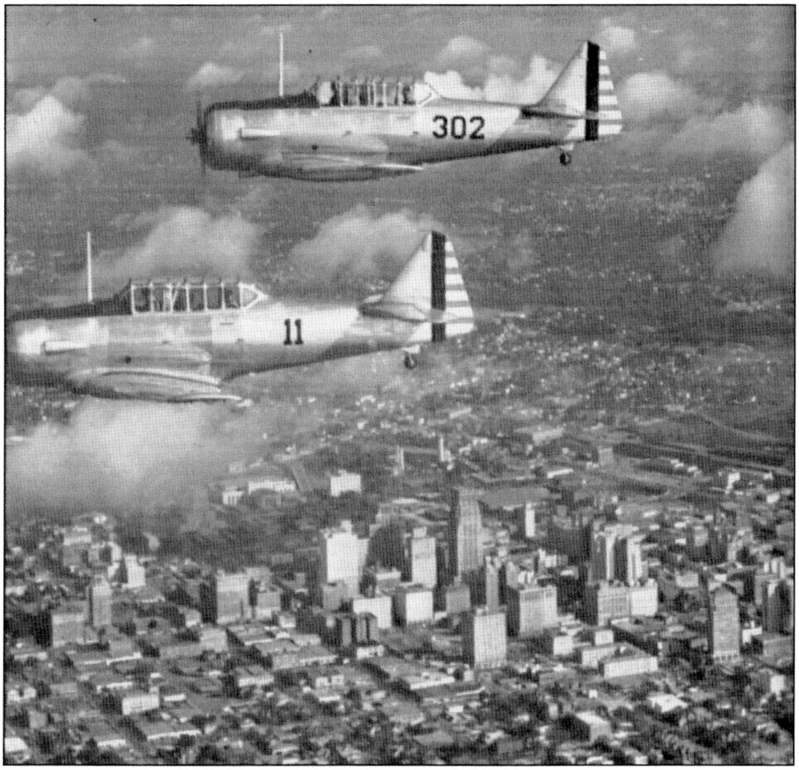

Two T-6 Texan advanced trainers from Ellington Field fly over Houston on a fine summer day during World War II. By 1940, Houston had grown from its 1920 population of 136,000 to 384,000 residents. (LSFM.)

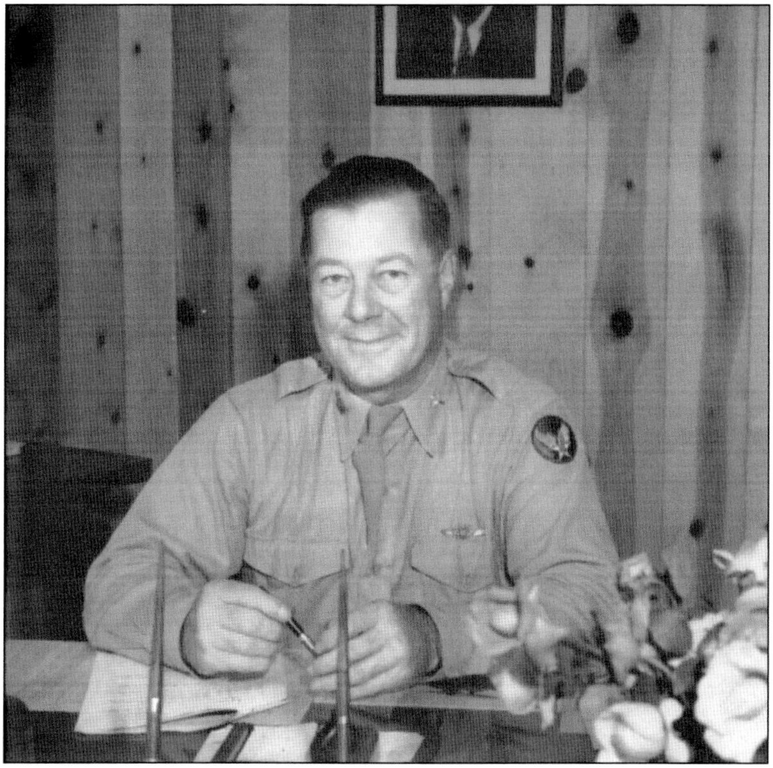

Ellington Field's commander when it reopened was Col. Walter Reid. He would command Ellington until April 1944. Reid was familiar with Ellington. He finished his flight training here in World War I. He received the Legion of Merit for his work running Ellington. After his retirement from the Army in 1949, he returned to Houston, eventually serving as a director in the city's aviation department. (LSFM.)

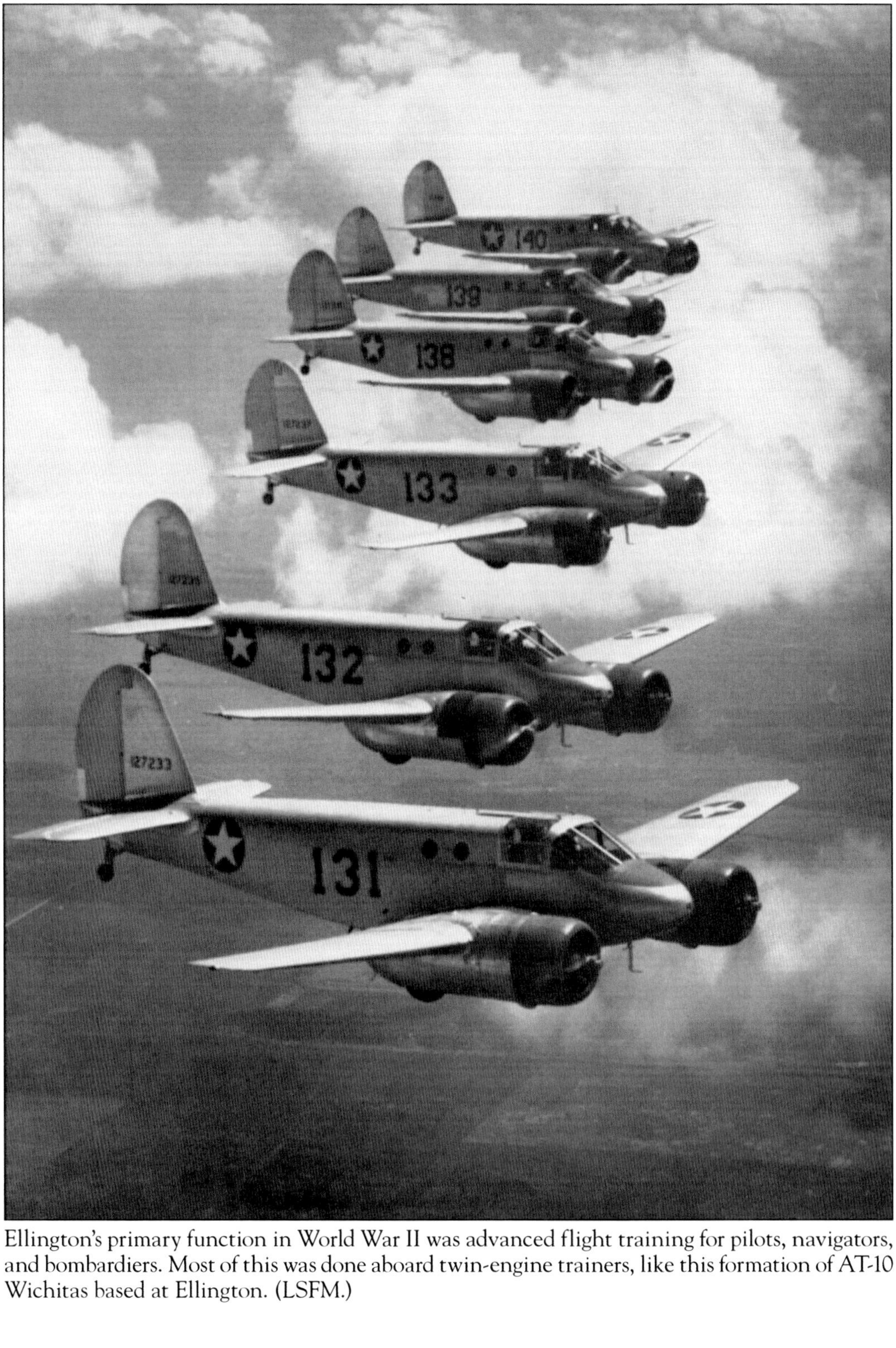

Ellington's primary function in World War II was advanced flight training for pilots, navigators, and bombardiers. Most of this was done aboard twin-engine trainers, like this formation of AT-10 Wichitas based at Ellington. (LSFM.)

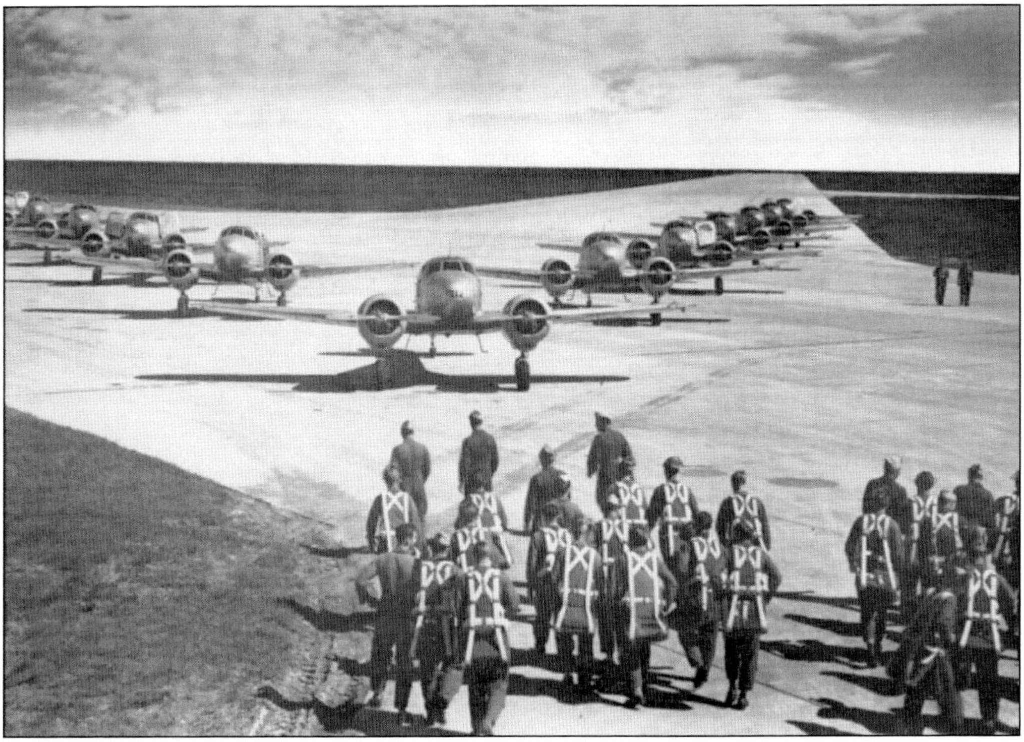

Student pilots move out to man their aircraft. This is advanced training on multi-engine aircraft. Not only are the aircraft more complex than the World War I trainers flown out of Ellington, so is the equipment worn by the students, including parachutes. This photograph was taken at the junction of Ellington's two main runways. (LSFM.)

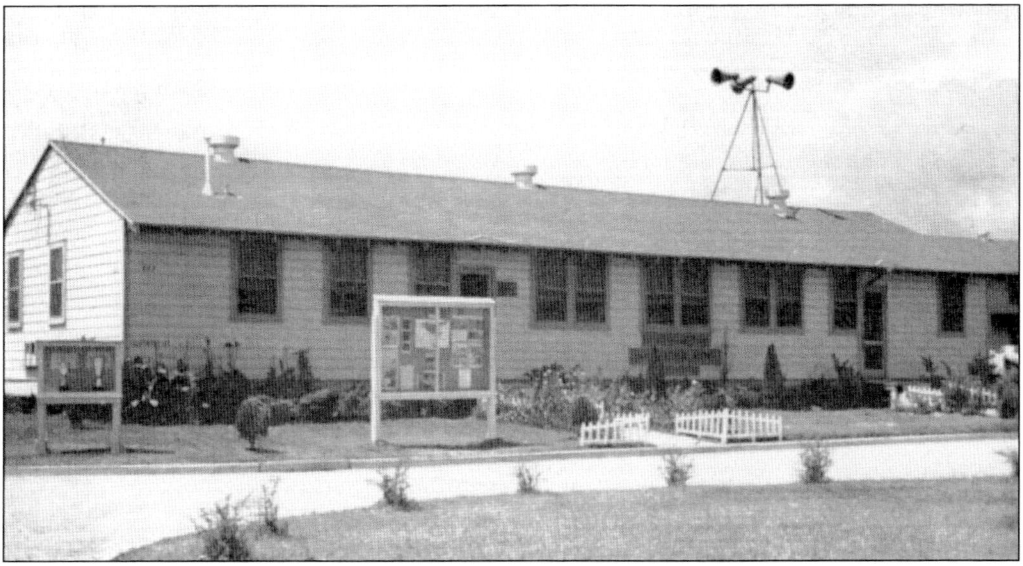

Headquarters for the Bombardier Training Wing at Ellington are seen here. This was one of 165 wooden buildings hastily erected in the winter of 1940–1941. Despite their temporary nature, some lasted at Ellington until the late 1980s. One, sold and moved off-site, became an astronaut hangout, the Outpost Tavern. It was finally torn down in 2007. (LSFM.)

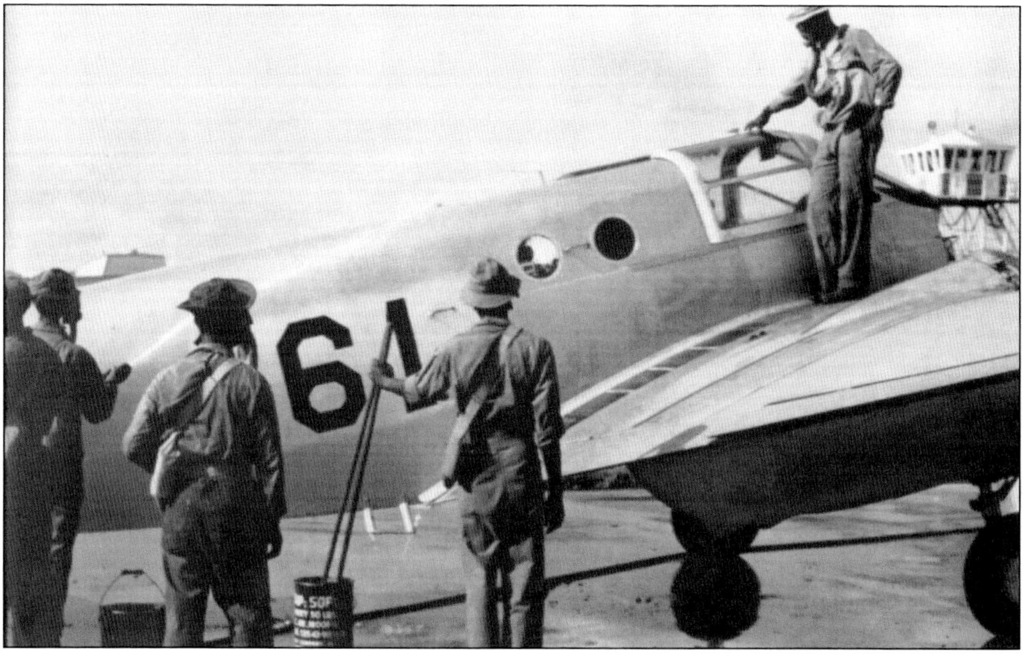

Flight training was not the only thing taught at Ellington Field; other fields of instruction were ground operations and maintenance. These ground crewmen are practicing chemical decontamination techniques on an AT-17. They are hosing it down and scrubbing it off to remove persistent chemical agents like mustard gas. (LSFM.)

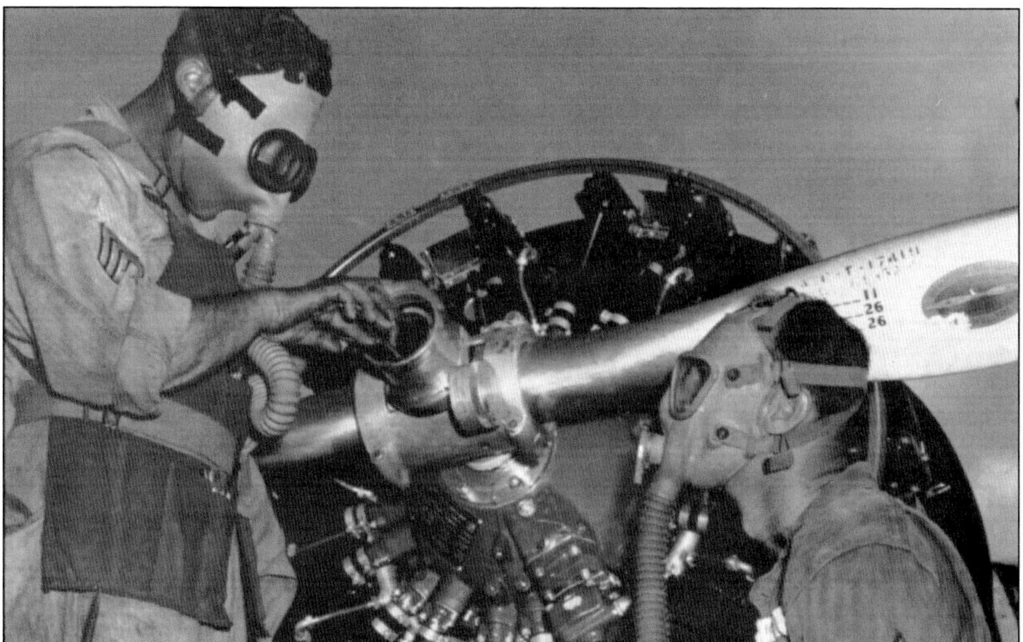

These two men practice decontaminating an engine. Although chemical agents were not used on the battlefield during World War II, no one was sure they would not be. Hence the training in preparing to defeat it. (LSFM.)

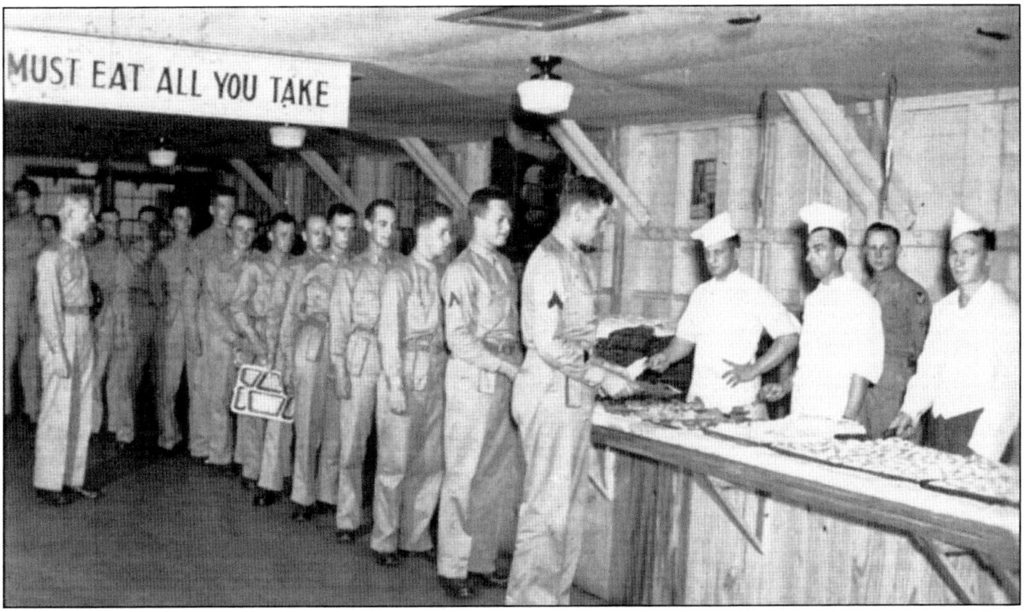

Mess was pretty much the same as in World War I too. Students would line up and go through the chow line. A soldier could have the mess clerks fill up his metal mess tray with as much food as he wanted—an unusual experience for many who grew up in the Great Depression. But as the sign stated, if you took it, you were expected to eat it. (LSFM.)

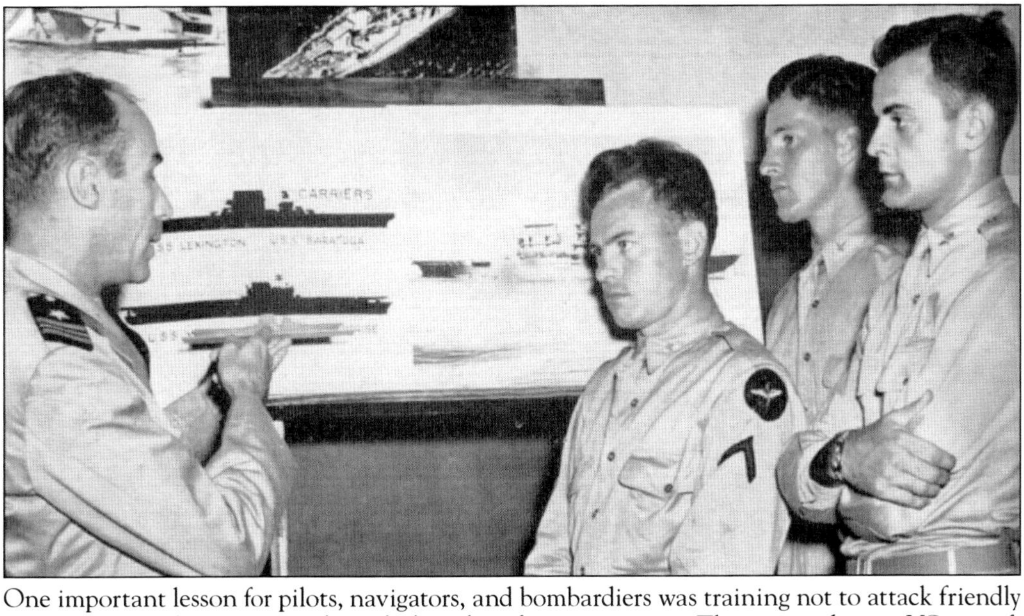

One important lesson for pilots, navigators, and bombardiers was training not to attack friendly forces. These cadets are going through ship identification training. The picture shows a US aircraft carrier. In action, they would get points taken off for attacking this ship. (LSFM.)

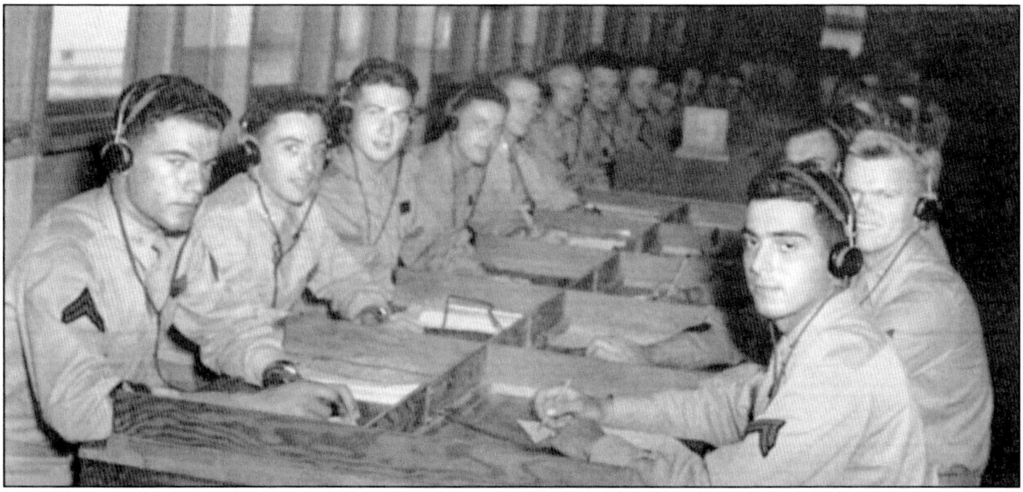

Radios, too, were more sophisticated than they had been in World War I and also more universal. These cadets are in class practicing reading Morse code messages. (LSFM.)

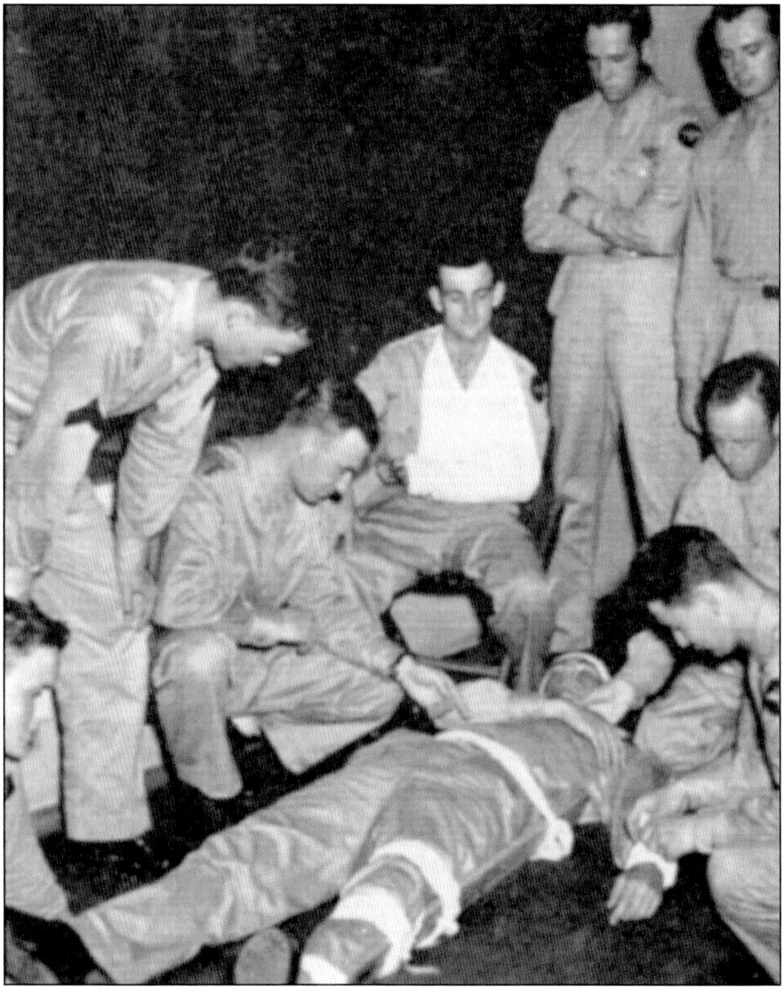

First aid was another skill critically needed by aircrew. If someone aboard an airplane were injured during a combat mission, a hospital might be hours away. In this first aid class, cadets learn how to keep their companions alive until they get home. (LSFM.)

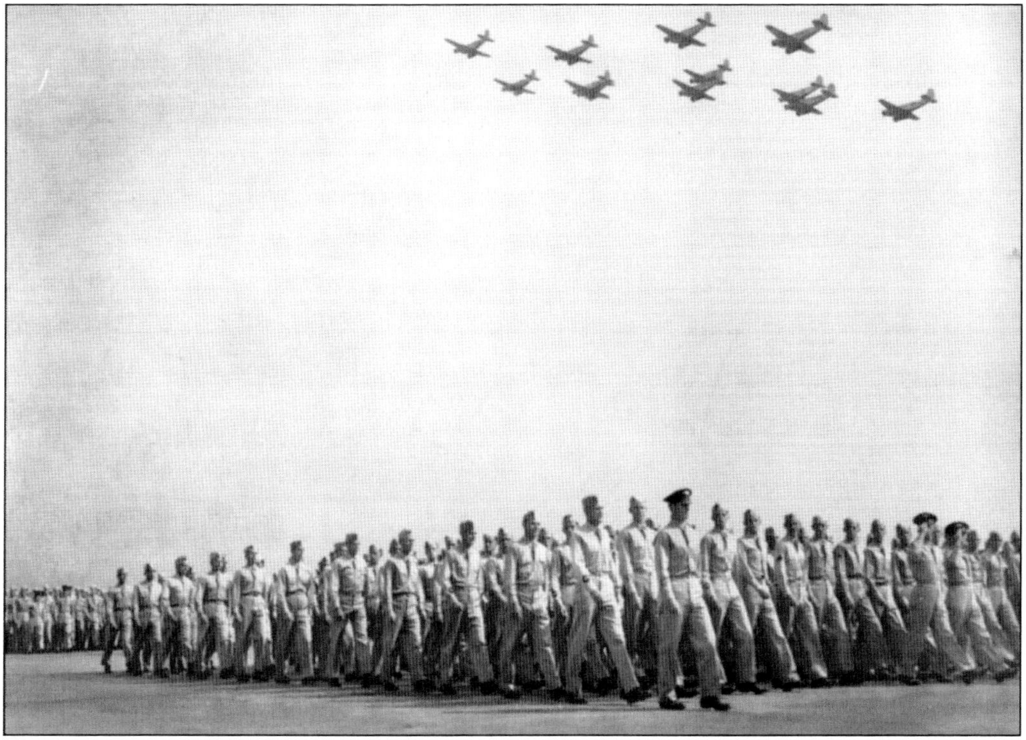

The graduating students, having completed their training, pass in review before the commanding officers as a formation of AT-17s fly overhead in an aerial salute. (LSFM.)

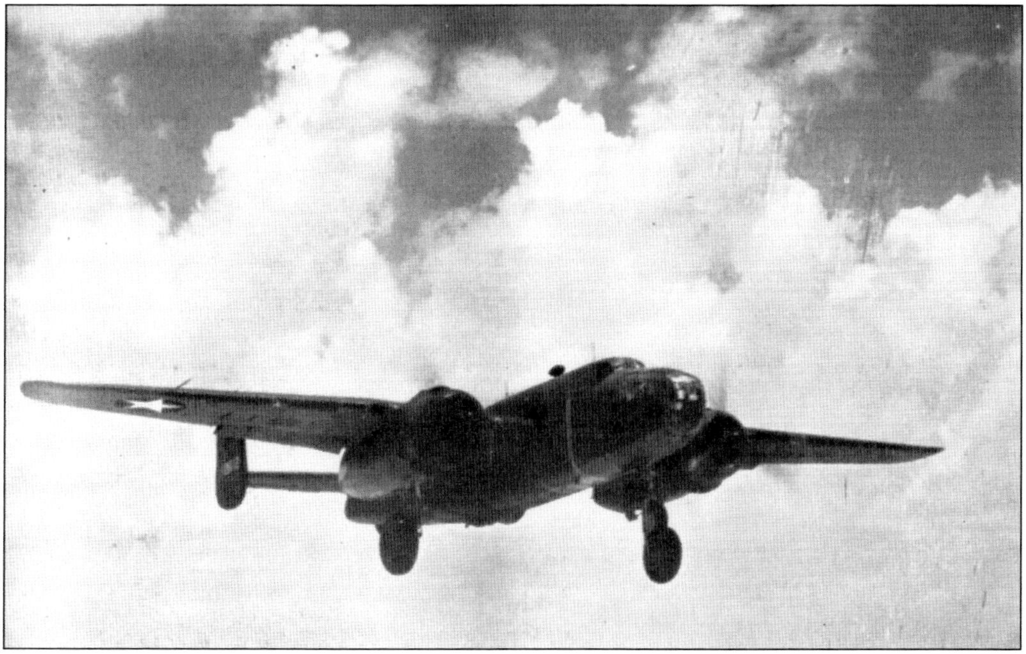

Graduation meant assignment to combat aircraft, typically multi-engine aircraft like this B-25 Mitchell, pictured landing at Ellington Field. While not stationed operationally at Ellington, aircraft like these stopped by to inspire the students. (LSFM.)

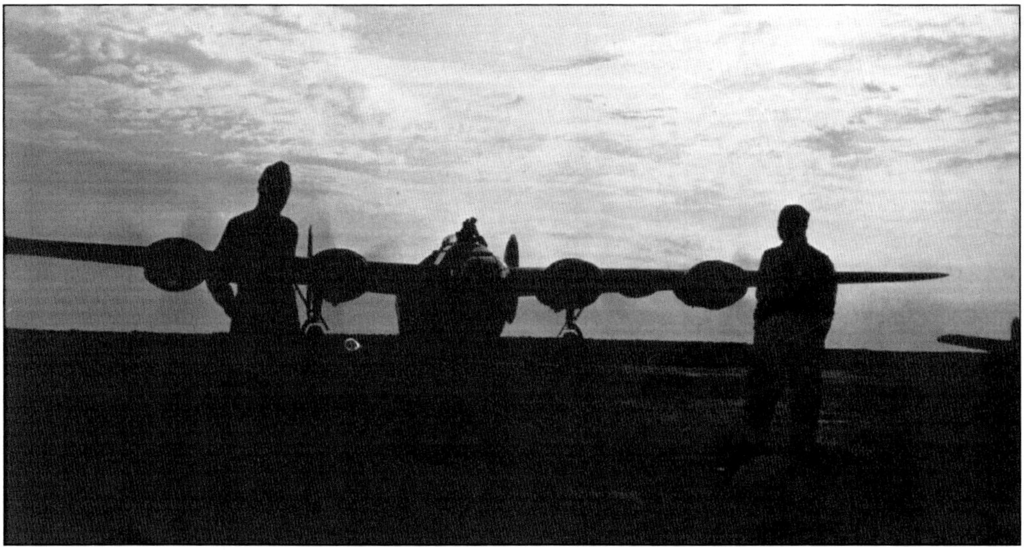

This included the big boys, like this B-24, arriving at Ellington at dusk, while two aviation cadets watch it taxi. They probably wondered when they would get their turn in one. (LSFM.)

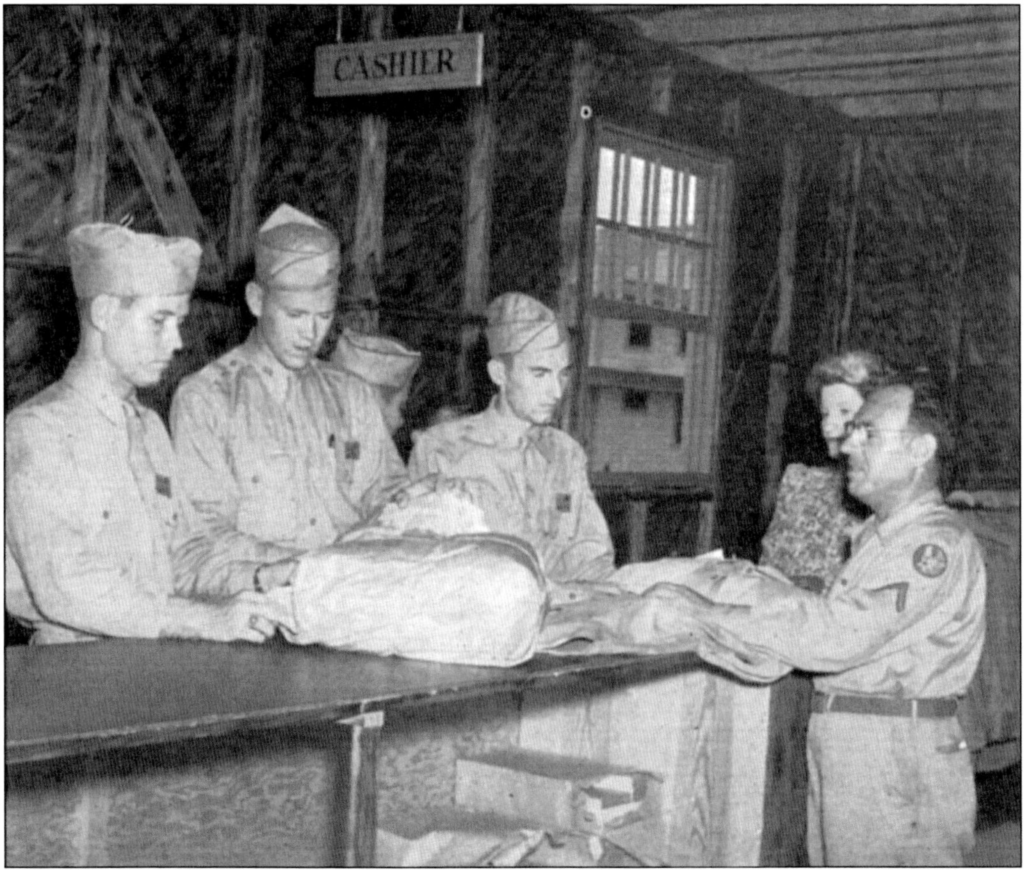

Life for students at Ellington in World War II was similar to what it was 23 years earlier. Cadets still went to the supply clerk to get their gear. (LSFM.)

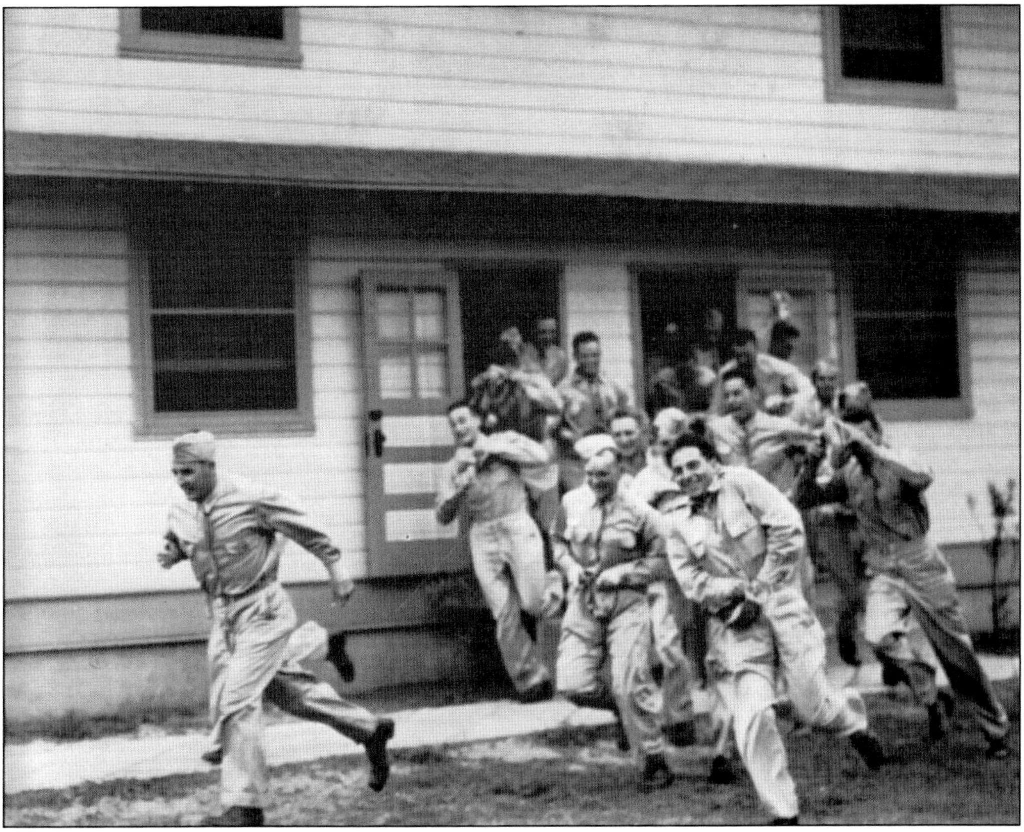

When "Reveille" sounded, cadets came out of the barracks on the double to fall in for morning formation. So as not to be late, they came out even if not fully dressed. Button that collar, tuck in that shirt, and cinch the belt on the way. (LSFM.)

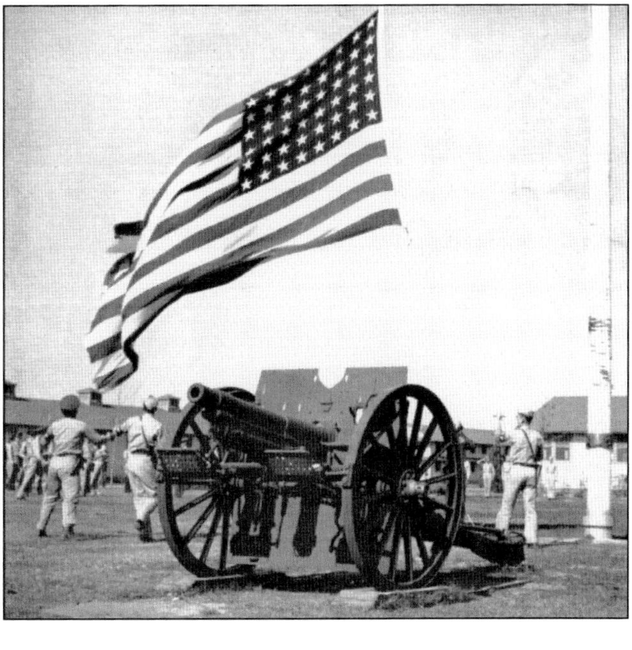

Everyone had better be in formation at attention with everything squared away when morning post sounded and the flag-raising ceremony began. There was even a howitzer for salutes, when necessary. (LSFM.)

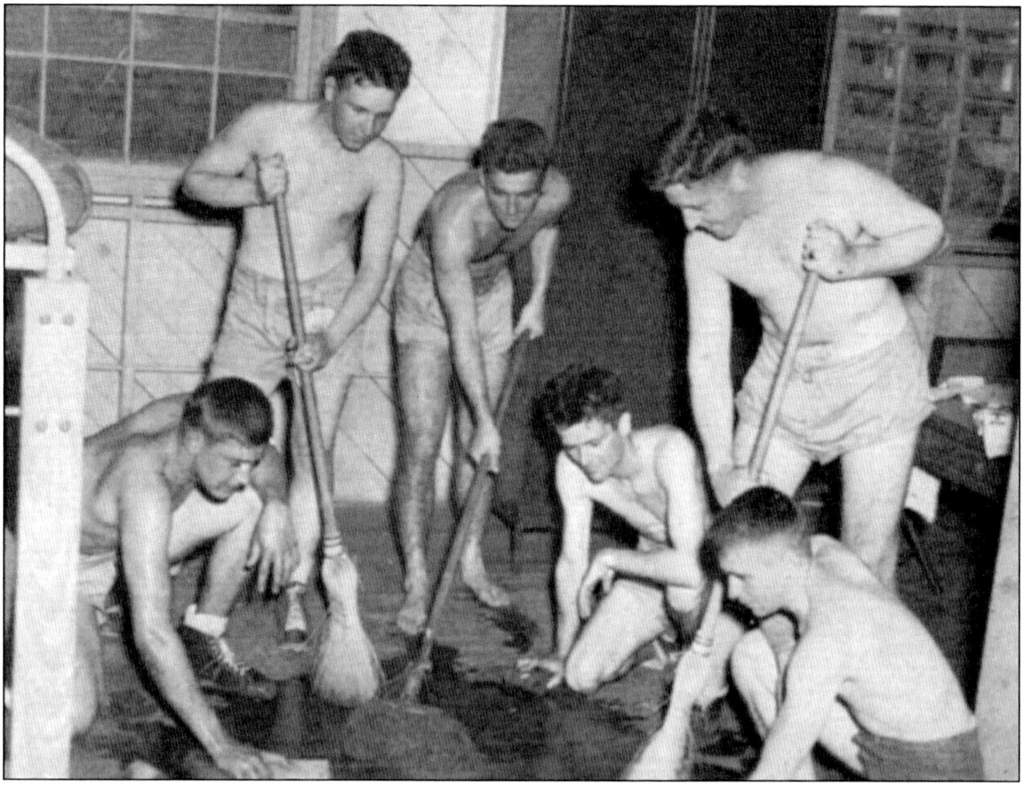

Life at Ellington also gave opportunities for an active social life with the opportunity to attend a lot of parties—as long as your taste for parties ran to GI parties. At these, your guests of honor were brooms, mops, sponges and elbow grease, as you prepared your barracks for inspection. (LSFM.)

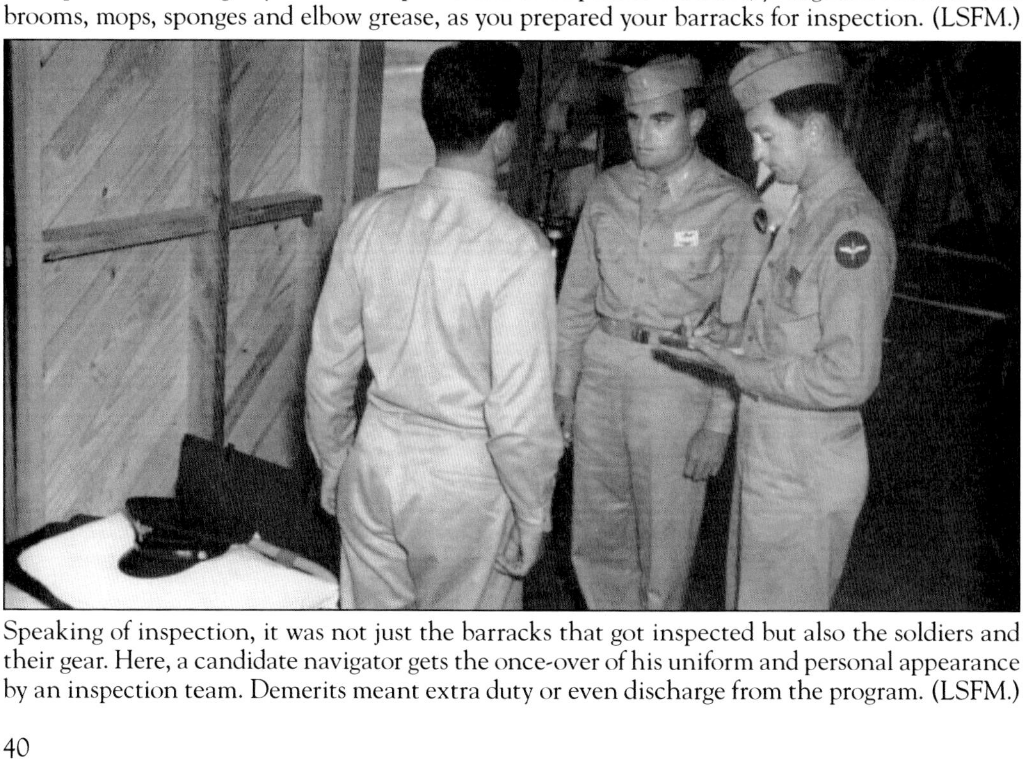

Speaking of inspection, it was not just the barracks that got inspected but also the soldiers and their gear. Here, a candidate navigator gets the once-over of his uniform and personal appearance by an inspection team. Demerits meant extra duty or even discharge from the program. (LSFM.)

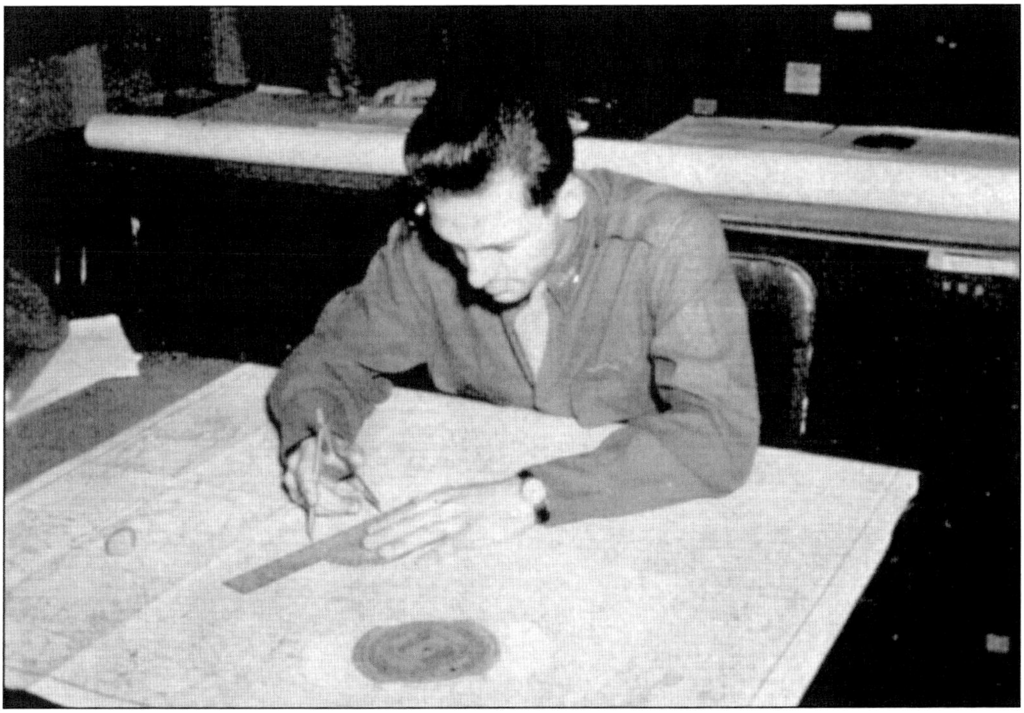

In addition to pilot training, Ellington trained many navigators for the Army Air Force during World War II. Navigation training quickly eclipsed bombardier training as Houston's cloudy weather interfered with bombardier training, while giving navigators a more challenging education. This navigation cadet practices map skills. (LSFM.)

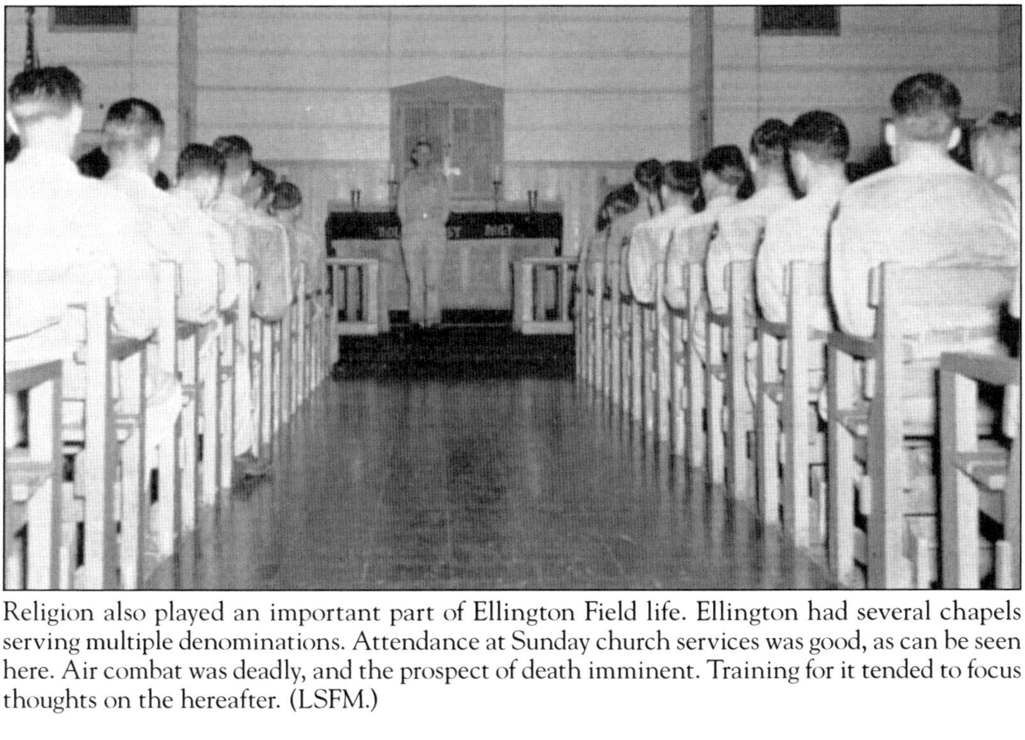

Religion also played an important part of Ellington Field life. Ellington had several chapels serving multiple denominations. Attendance at Sunday church services was good, as can be seen here. Air combat was deadly, and the prospect of death imminent. Training for it tended to focus thoughts on the hereafter. (LSFM.)

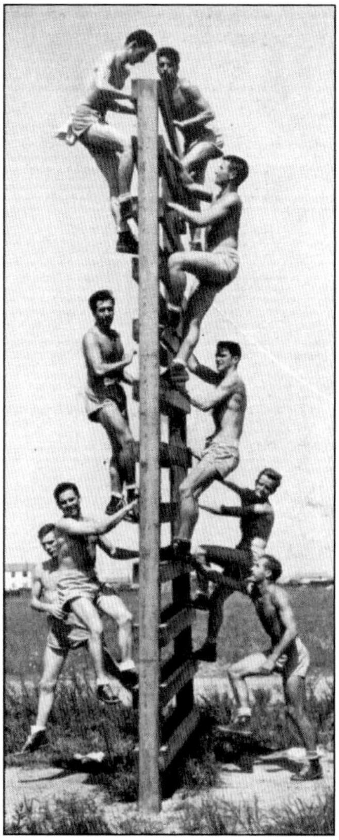

Athletics played an important part at Ellington as in the last war. Good physical conditioning was critical to aircrews' successful performance. Here, aviation cadets pose on a climbing wall. (LSFM.)

Boxing remained as popular in World War II as it had been a generation earlier. These two soldiers mix it up, sparring behind the barracks. (LSFM.)

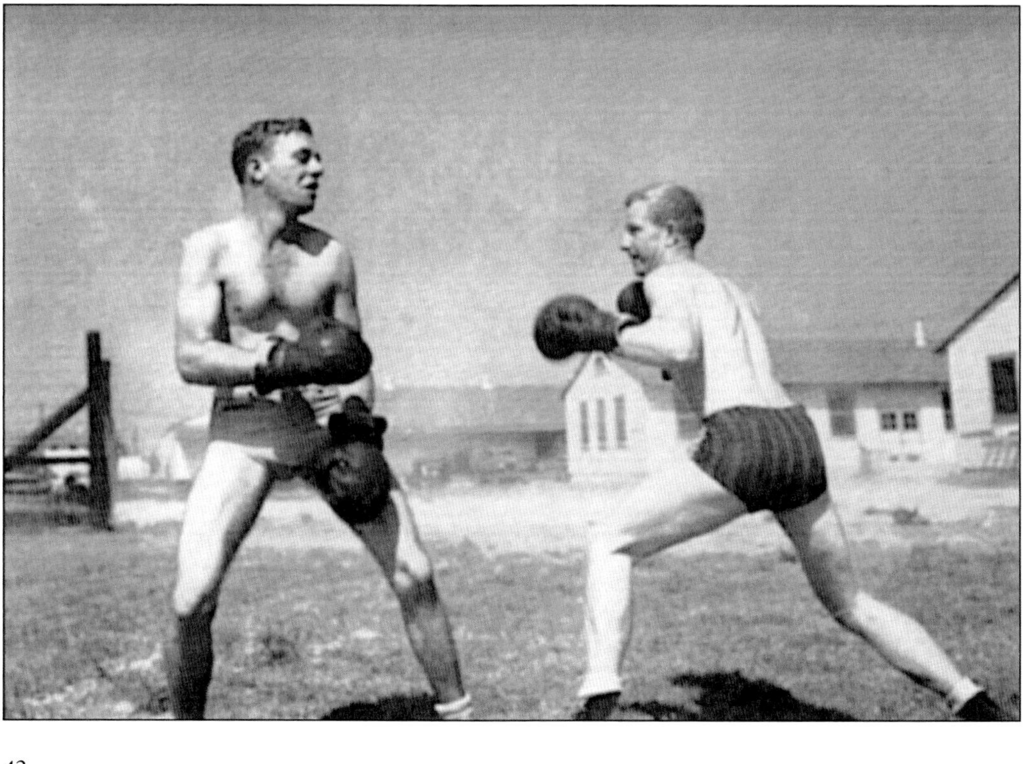

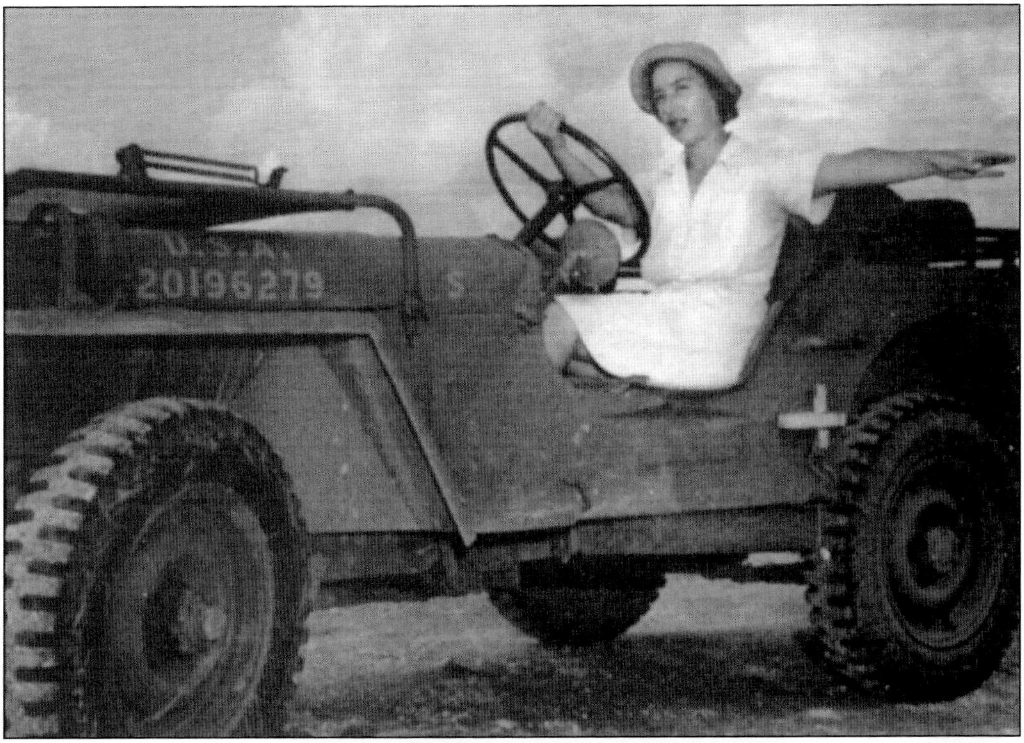

The Women's Auxiliary Corps was formed during World War II to allow women to serve in non-combatant roles. Many WACs served at Ellington, including this woman driving a jeep. Her white uniform indicates she is a nurse. Unlike her World War I counterparts, she had a rank and was considered a soldier, not a civilian. (LSFM.)

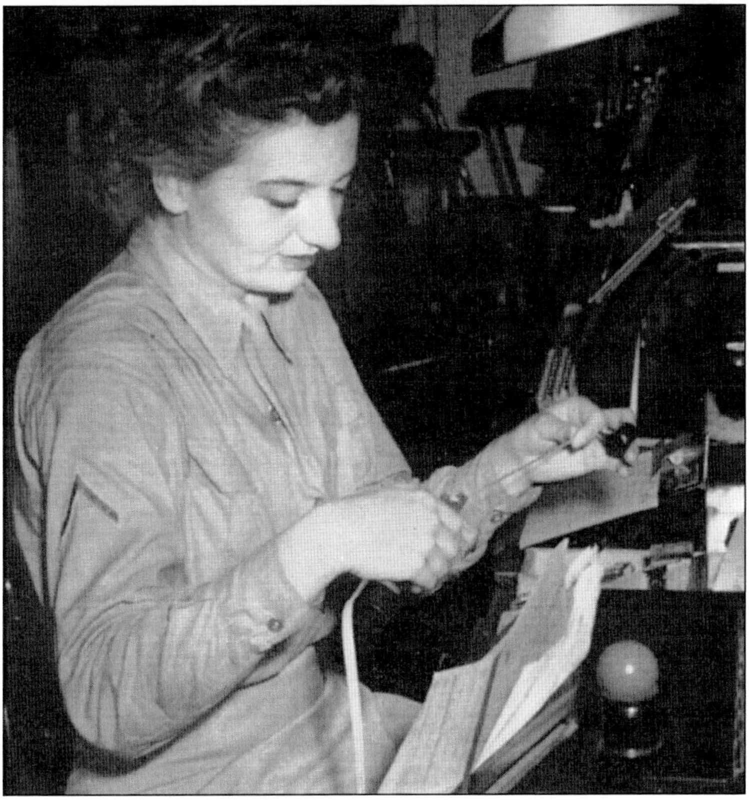

Women also served as clerks and in other non-combat roles, typically in offices, like this WAC private. Each desk job taken by a woman freed up a man to serve in a combat role. (LSFM.)

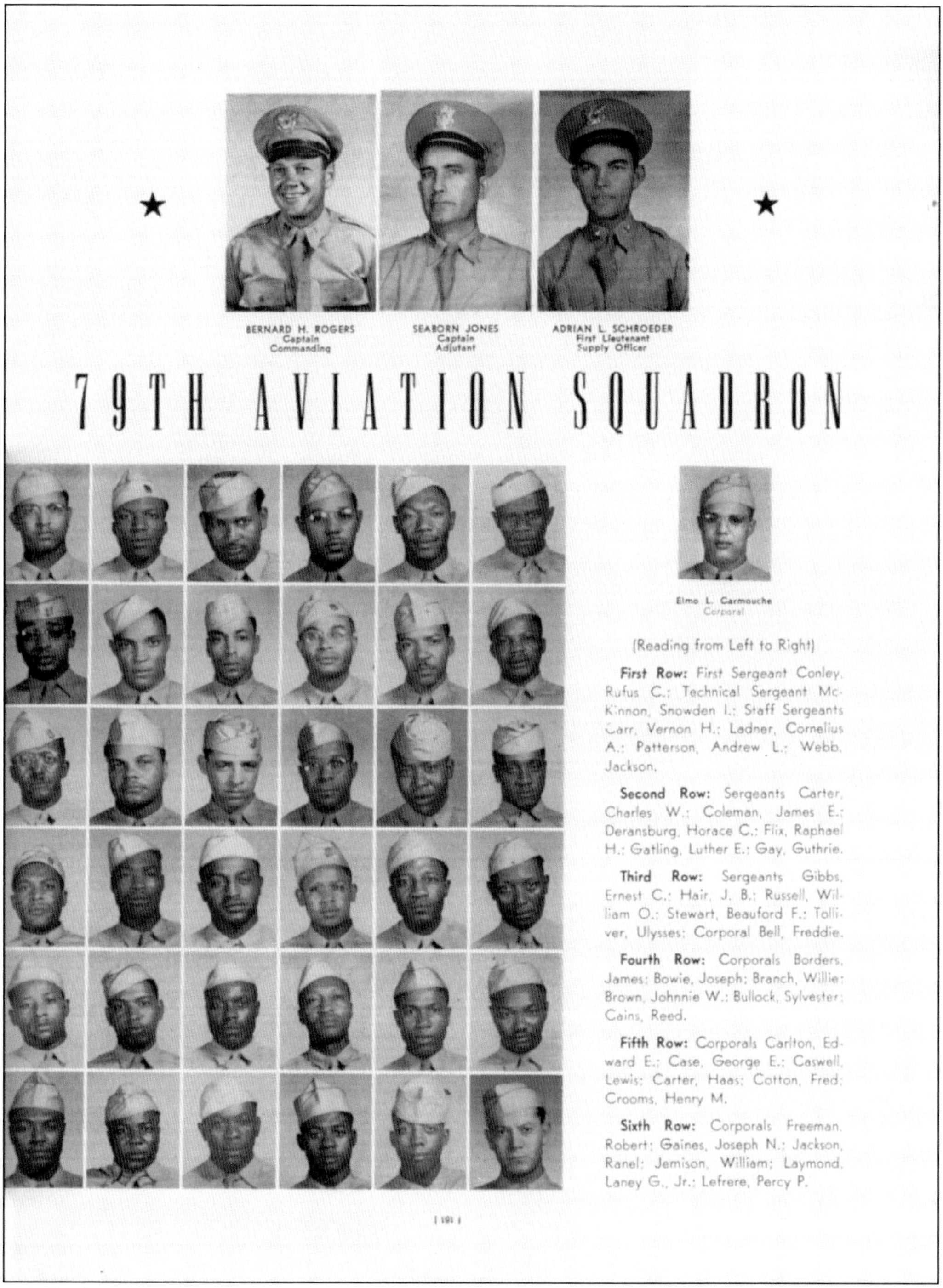
BERNARD H. ROGERS
Captain
Commanding

SEABORN JONES
Captain
Adjutant

ADRIAN L. SCHROEDER
First Lieutenant
Supply Officer

79TH AVIATION SQUADRON

Elmo L. Carmouche
Corporal

(Reading from Left to Right)

First Row: First Sergeant Conley, Rufus C.; Technical Sergeant McKinnon, Snowden I.; Staff Sergeants Carr, Vernon H.; Ladner, Cornelius A.; Patterson, Andrew L.; Webb, Jackson.

Second Row: Sergeants Carter, Charles W.; Coleman, James E.; Deransburg, Horace C.; Flix, Raphael H.; Gatling, Luther E.; Gay, Guthrie.

Third Row: Sergeants Gibbs, Ernest C.; Hair, J. B.; Russell, William O.; Stewart, Beauford F.; Tolliver, Ulysses; Corporal Bell, Freddie.

Fourth Row: Corporals Borders, James; Bowie, Joseph; Branch, Willie; Brown, Johnnie W.; Bullock, Sylvester; Cains, Reed.

Fifth Row: Corporals Carlton, Edward E.; Case, George E.; Caswell, Lewis; Carter, Haas; Cotton, Fred; Crooms, Henry M.

Sixth Row: Corporals Freeman, Robert; Gaines, Joseph N.; Jackson, Ranel; Jemison, William; Laymond, Laney G., Jr.; Lefrere, Percy P.

[191]

While the US Army was still segregated in World War II, in that war, unlike World War I, African Americans were allowed to serve on aircraft as pilots and aircrew, albeit in segregated units. At Ellington, that unit was the 79th Aviation Squadron, shown here. This page shows the men from one class. (LSFM.)

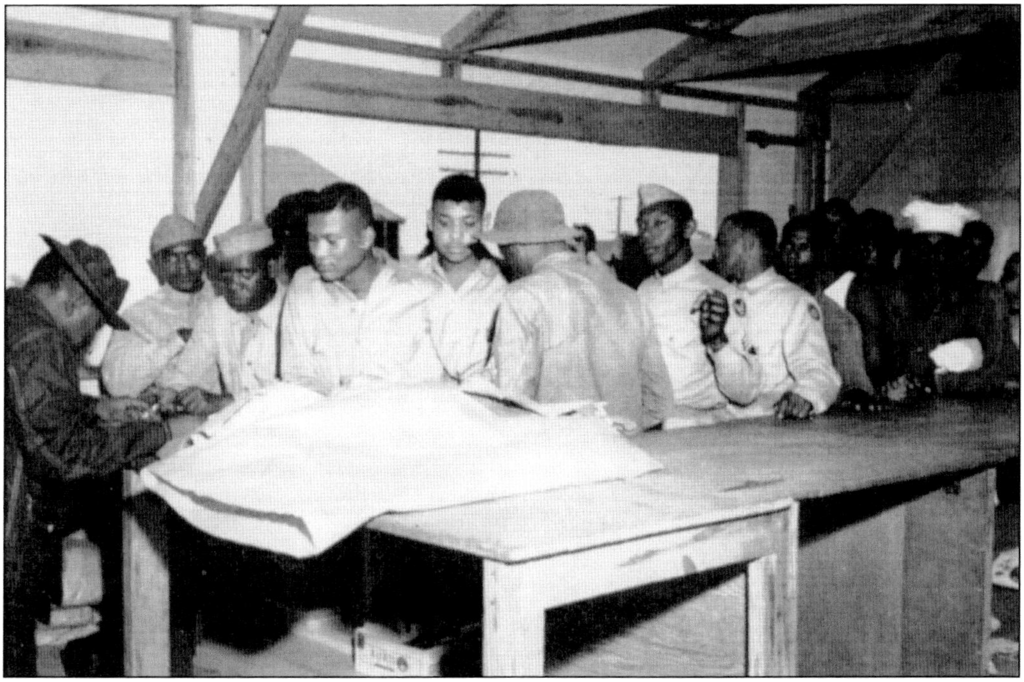

Despite the color barriers, the lives of the men serving in the 79th Aviation Squadron were similar to those of their white counterparts. Here, a bunch of aviation candidates go through a supply line getting their gear, preparatory to starting training. (LSFM.)

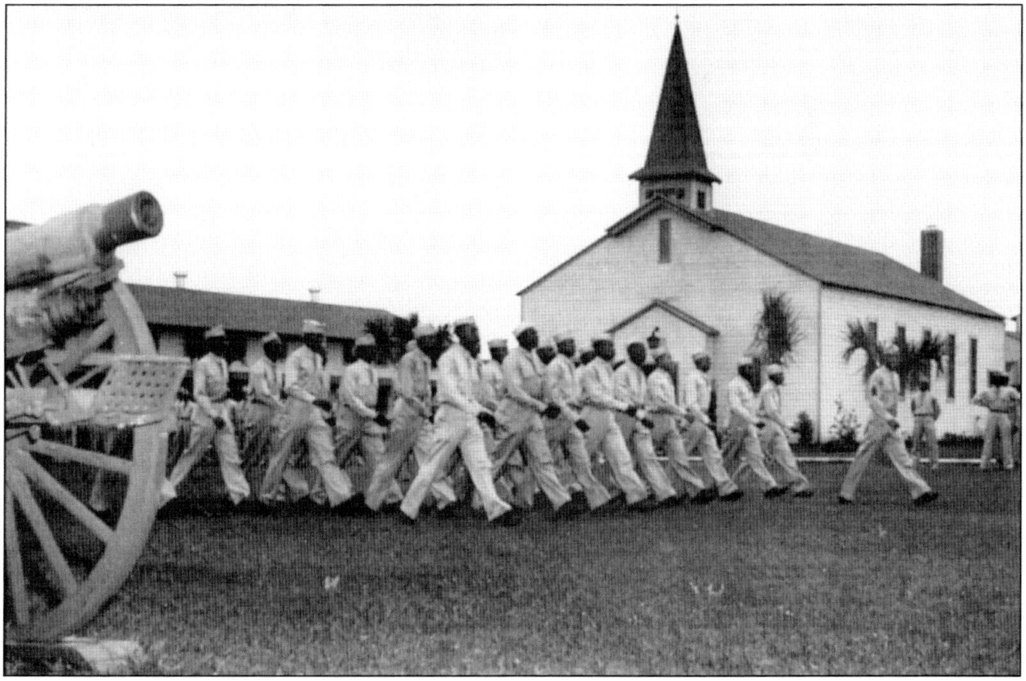

These 79th Squadron soldiers march through the parade ground in formation. They worked hard to look sharp and present a military appearance, realizing they had to perform twice as well as the white troops to receive half the credit for their work. (LSFM.)

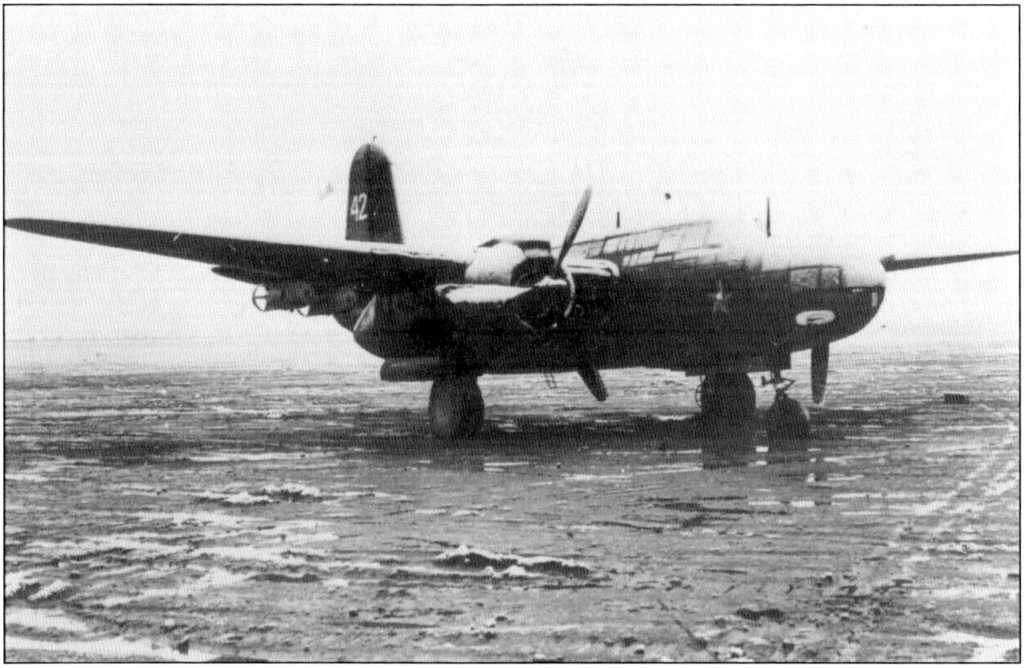

As for the 111th Observation Squadron, following its brief stay at Ellington, it was sent across the Atlantic as part of the 68th Observation Group doing antisubmarine patrols. This photograph was taken in early 1943 at Oujda, French Morocco, in North Africa when the unit was flying A-20 Havocs. (NMUSAF.)

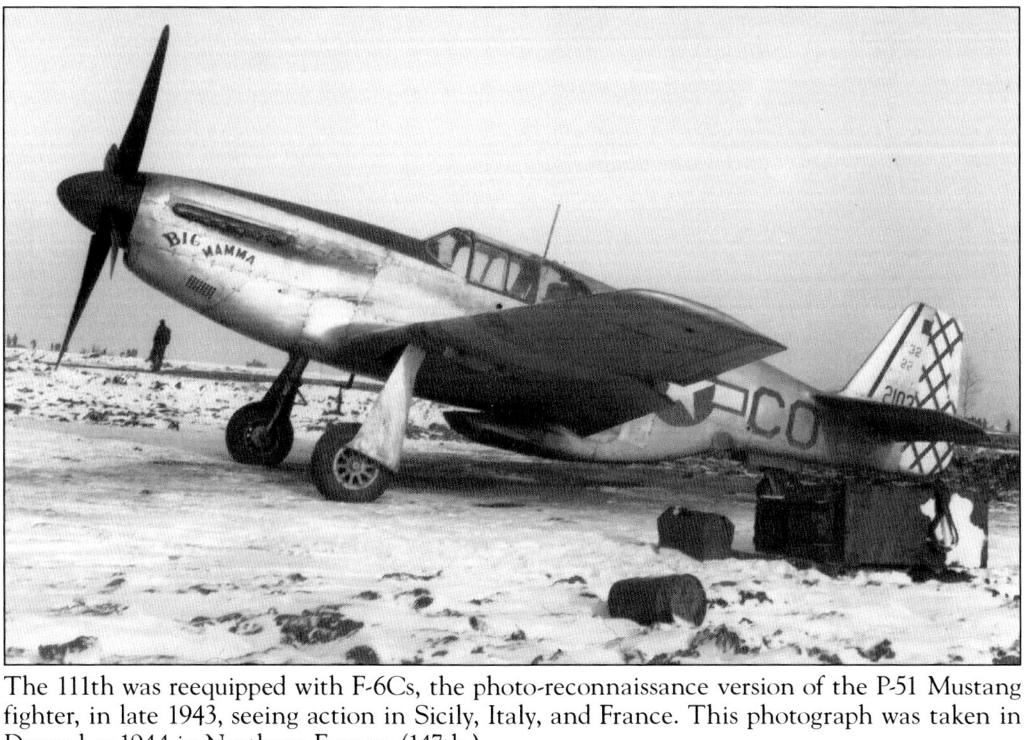

The 111th was reequipped with F-6Cs, the photo-reconnaissance version of the P-51 Mustang fighter, in late 1943, seeing action in Sicily, Italy, and France. This photograph was taken in December 1944 in Northern France. (147th.)

Three

Cold Warrior 1947–1984

After World War II ended, Ellington Field closed for a second time, deactivating on March 11, 1946. As a result of a new war, Ellington reopened less than a year later, in February 1947. This was the opening of the Cold War, a 45-year conflict between the United States and the Soviet Union that lasted until the latter's collapse in 1991.

Ellington served as an important training base in the 1950s and 1960s, but it took on a new role in the mid-1950s, one it kept through 1976. It became an integral part of the North American air defense system, providing a base for fighters shielding the Texas Gulf Coast from aerial attack. High-performance supersonic jets armed with air-to-air missiles patrolled the Gulf of Mexico. Sometimes during periods of tension, such as the Cuban Missile Crisis, their missiles were nuclear-tipped and the patrols continuous.

Through much of this period, Ellington served as the home of the 111th Fighter Squadron, part of the Texas National Guard, stationed in Houston. It flew F-51D Mustangs. Over the next 75 years, the 111th Squadron, in a number of different roles, used Ellington as its home base. The Mustangs were replaced by F-80 Shooting Stars, and then by F-86 Sabres, F-102 Delta Darts, F-101 Voodoos, and F-4 Phantoms. One fighter jockey piloting F-102s out of Ellington, Lt. George W. Bush, went on to become president of the United States.

Ellington also changed its name. With the creation of the United States Air Force in September 1947, all Army Air Force personnel, aircraft, and facilities were transferred to the new service. Ellington Field became Ellington Air Force Base. It kept that name until 1976, when the Air Force turned the base over to the Texas Air National Guard and it reverted to Ellington Field.

Over the next 35 years, Ellington served an increasingly important role in the defense of the Texas Gulf Coast.

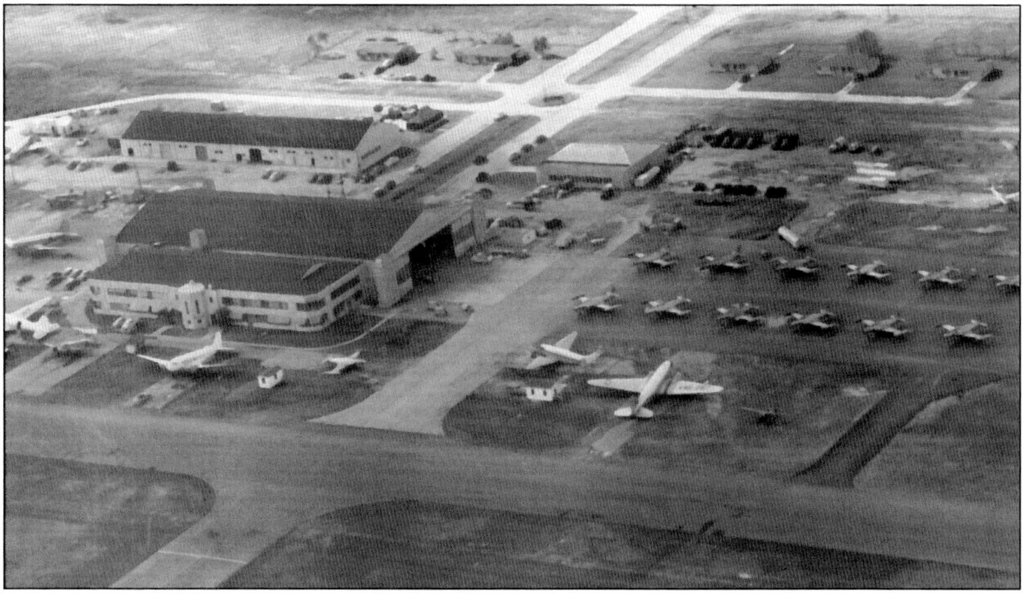

The 111th returned to Houston after World War II, reverting to its status as part of the Texas Air National Guard. While it would spend most of the Cold War at Ellington Air Force Base, when it first returned, Ellington was closed. As a result, the 111th was initially stationed at Houston Municipal Airport (today's Hobby Airport). (147th.)

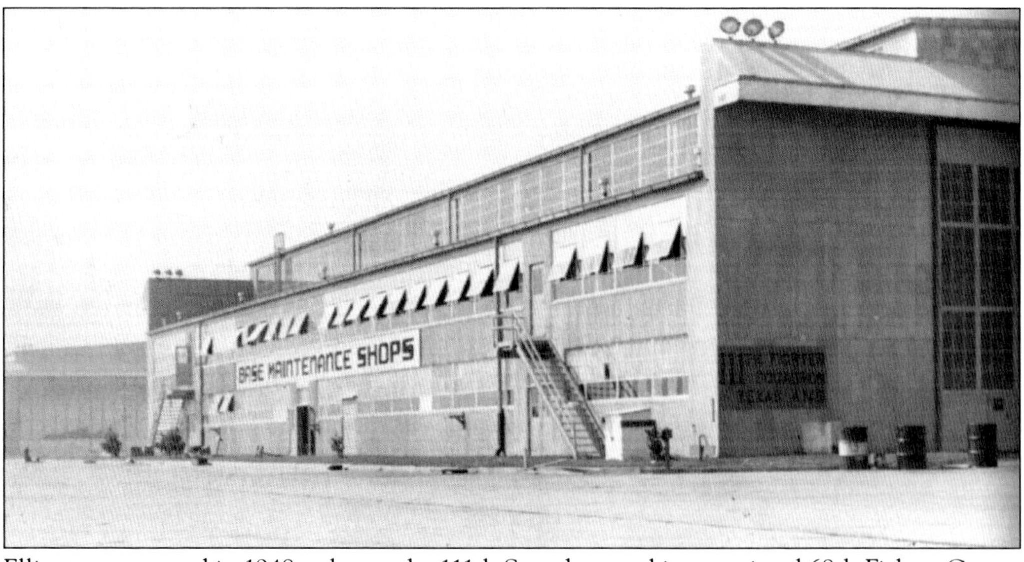

Ellington reopened in 1948 to house the 111th Squadron and its associated 68th Fighter Group. They operated out of this hangar until they were nationalized for the Korean War in 1950. They returned to Ellington Air Force Base in 1956 after a short exile at La Porte Airport. (Missy Bryant.)

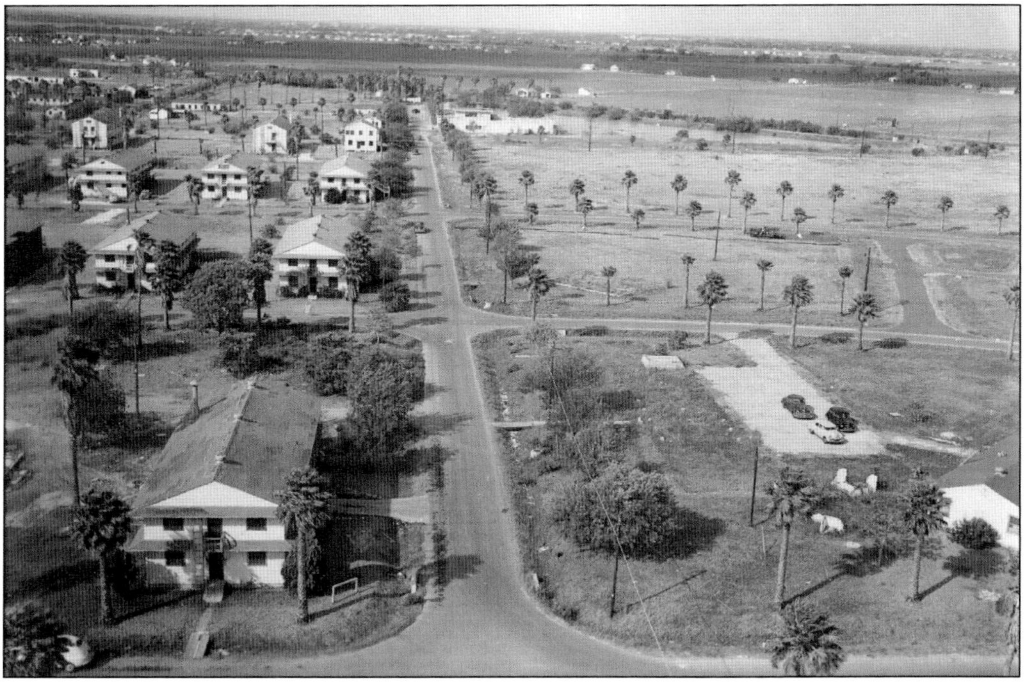

Ellington also opened a navigation training school in 1949. The barracks and office buildings erected in World War II offered students plenty of space to both live and learn. There was a little less hands-on navigation in the 1950s, however. Most of the course involved polar navigation, which was the route to and from the Soviet Union. (LSFM.)

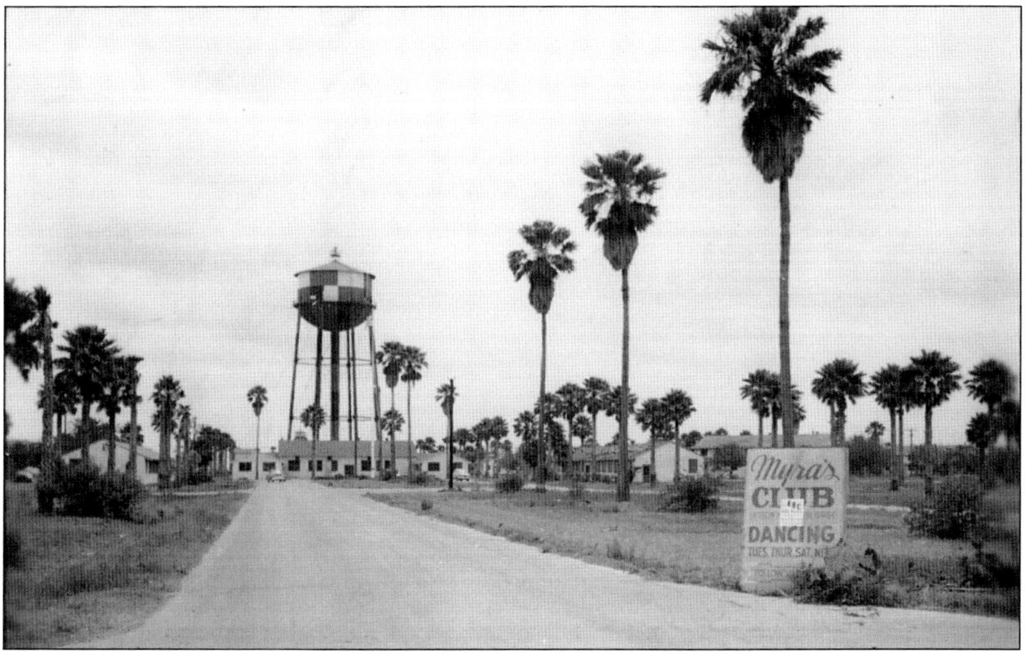

The main entrance to Ellington Air Force Base is seen here in the early 1950s. The Myra Club was at the American Legion post across Texas State Highway 3 from Ellington. (LSFM.)

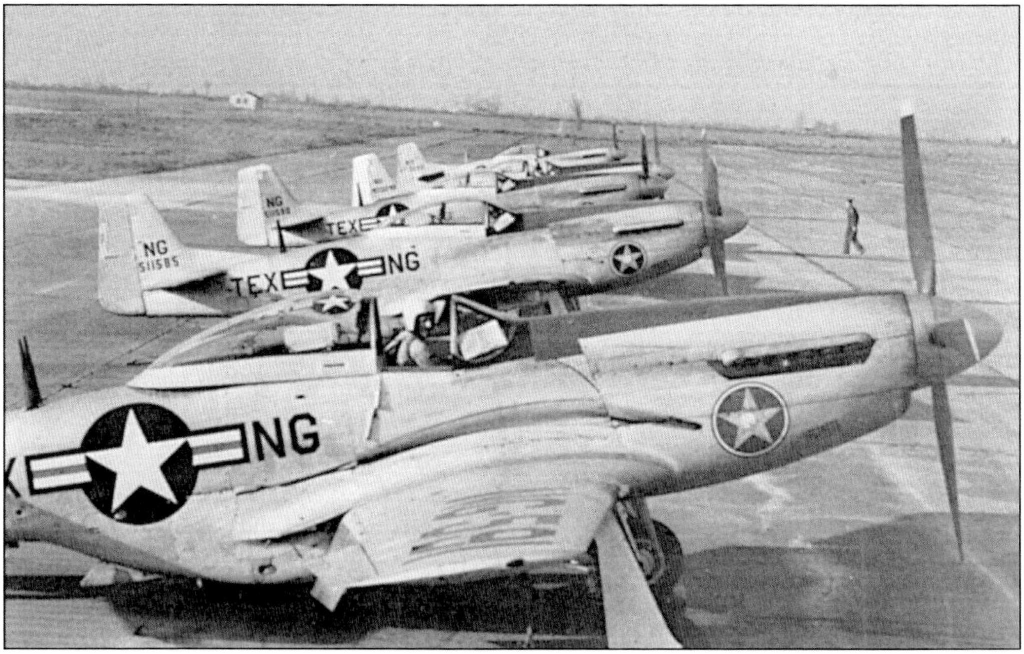

The 111th Squadron was operating F-51Ds in 1949. (The "P" in P-51—for pursuit—had been replaced by "F" for fighter in 1947.) The 111th's aircraft were used as fighter-bombers at the time. These F-51s, with the 111th's "Ace-in-the-Hole" unit insignia on their nose, are lined up on the apron at Ellington. (Missy Bryant.)

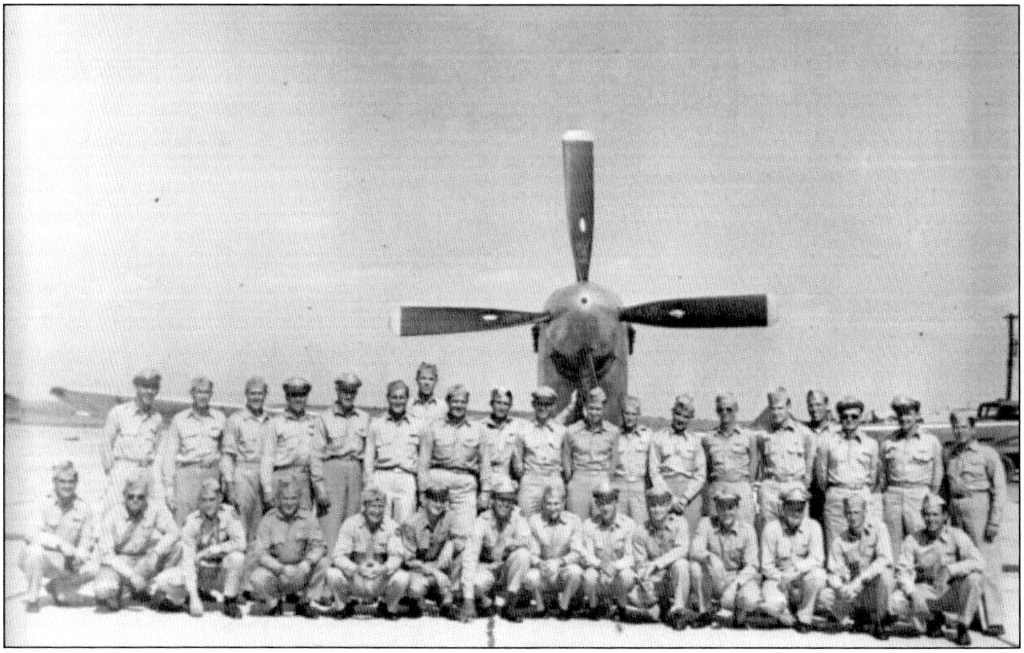

The 111th was an Air National Guard unit. Its members were volunteers who served weekends and in a two-week summer camp because they enjoyed flying and serving their country while also getting on with their lives. These were the officers of the 111th in 1949, posing in front of one of their aircraft. (Missy Bryant.)

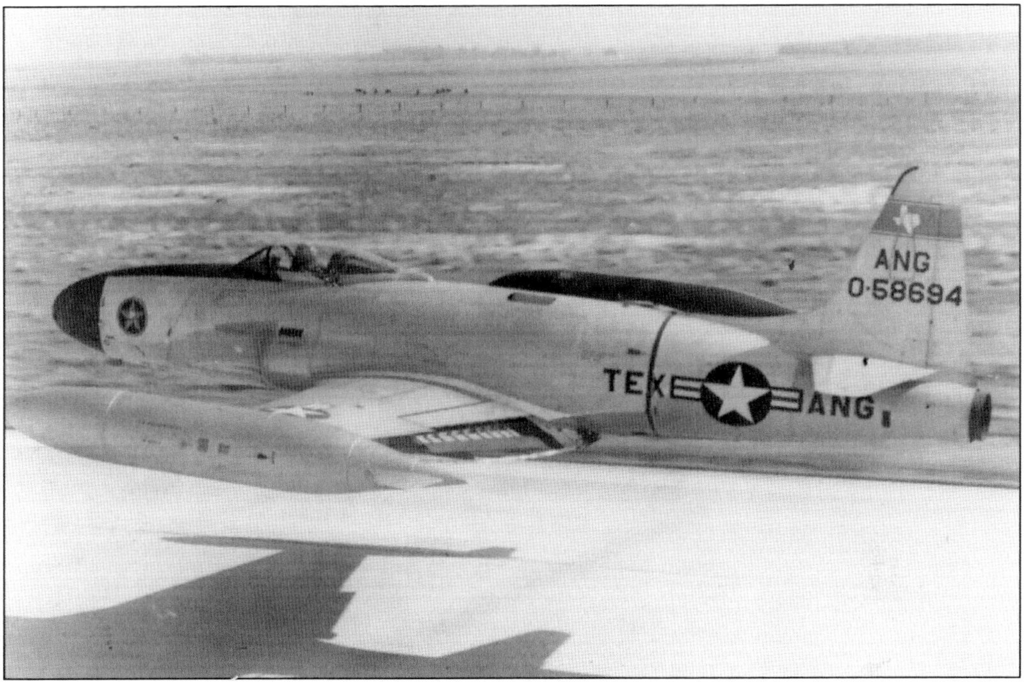

When the 111th was in Korea, it flew the F-80 Shooting Star. While it reverted to F-51s after returning to the states, it resumed flying F-80s in 1956 after returning to Ellington. This Ace-in-the-Hole F-80 is making a low pass over Ellington's runway. (147th.)

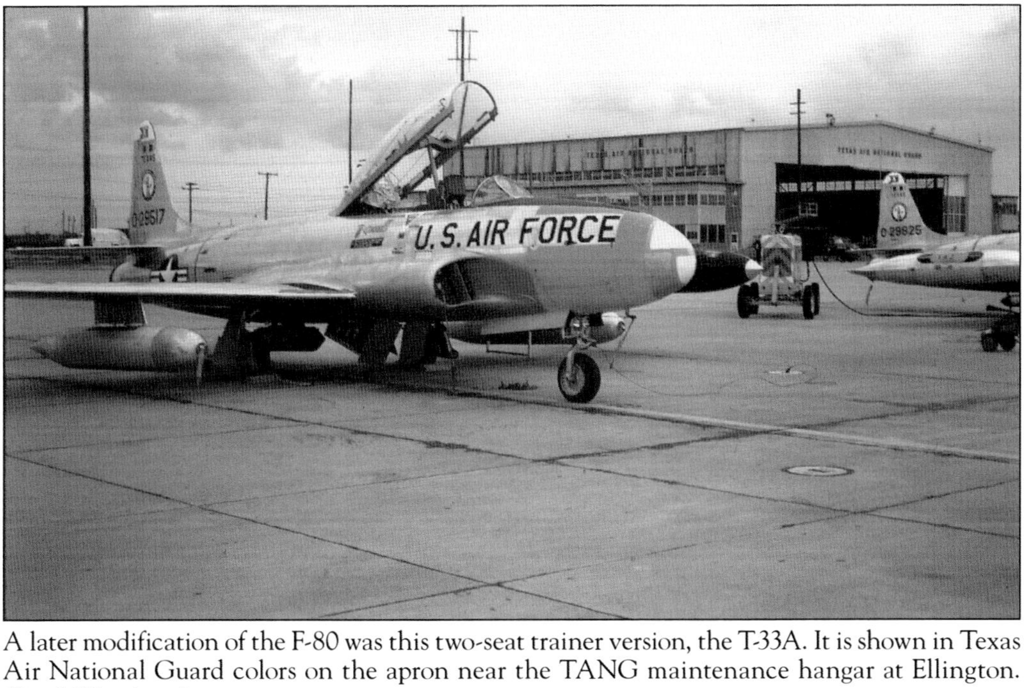

A later modification of the F-80 was this two-seat trainer version, the T-33A. It is shown in Texas Air National Guard colors on the apron near the TANG maintenance hangar at Ellington. (Boyd Waechter.)

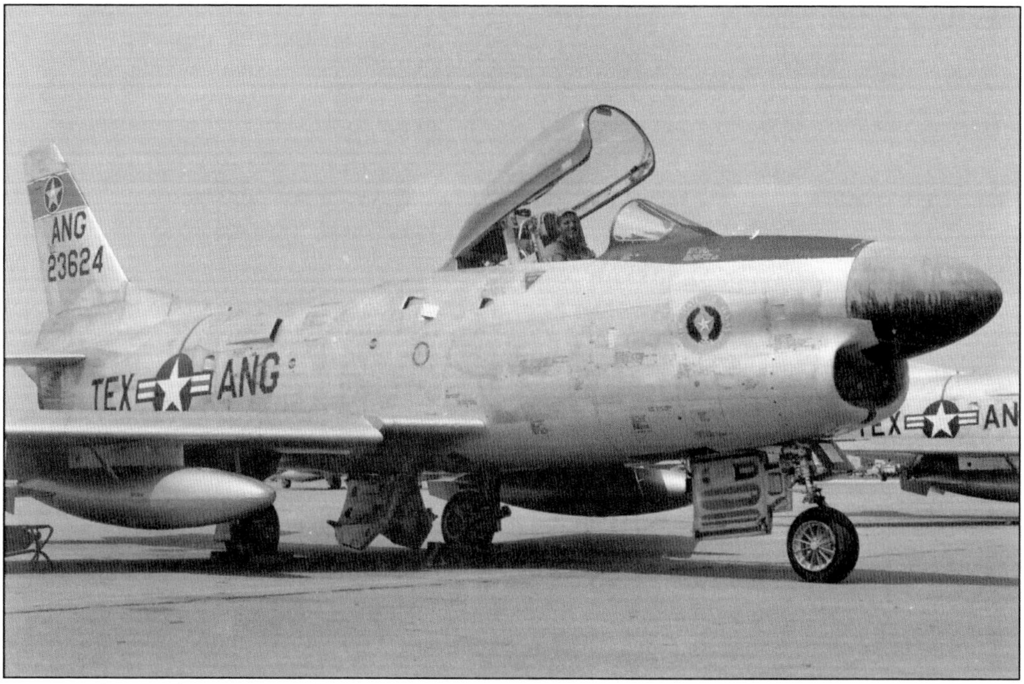

In 1957, the 111th re-equipped and flew the F-86D Saber at Ellington. This was a radar-equipped transonic version of the Korean War–era fighter, capable of operating in all weather and at night. It better suited the around-the-clock air defense mission Ellington was then filling. (147th.)

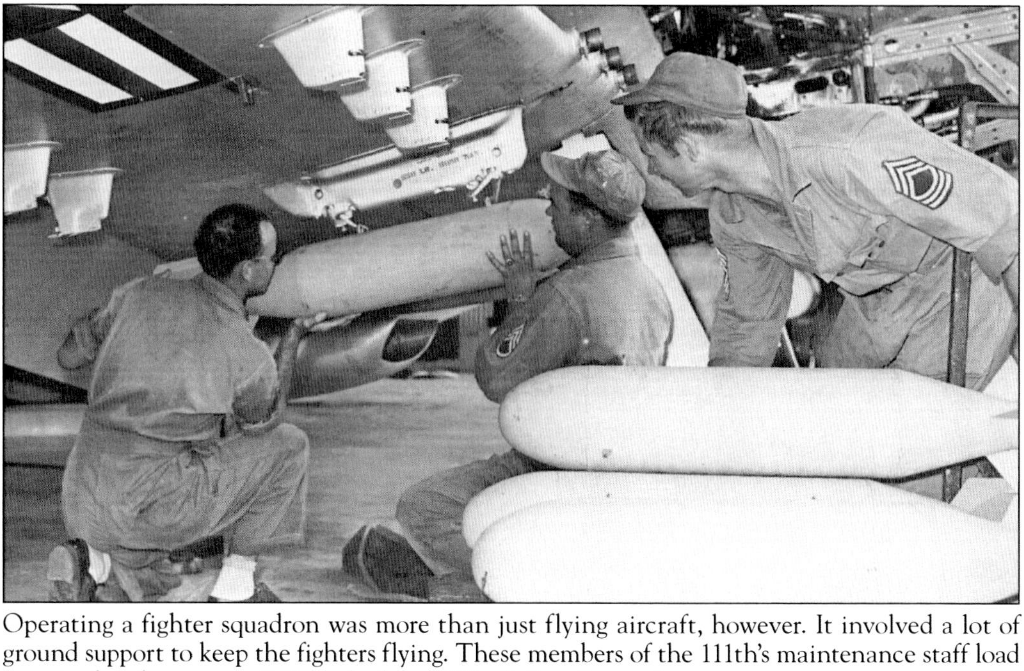

Operating a fighter squadron was more than just flying aircraft, however. It involved a lot of ground support to keep the fighters flying. These members of the 111th's maintenance staff load practice bombs onto an F-51 in the late 1940s. (Missy Bryant.)

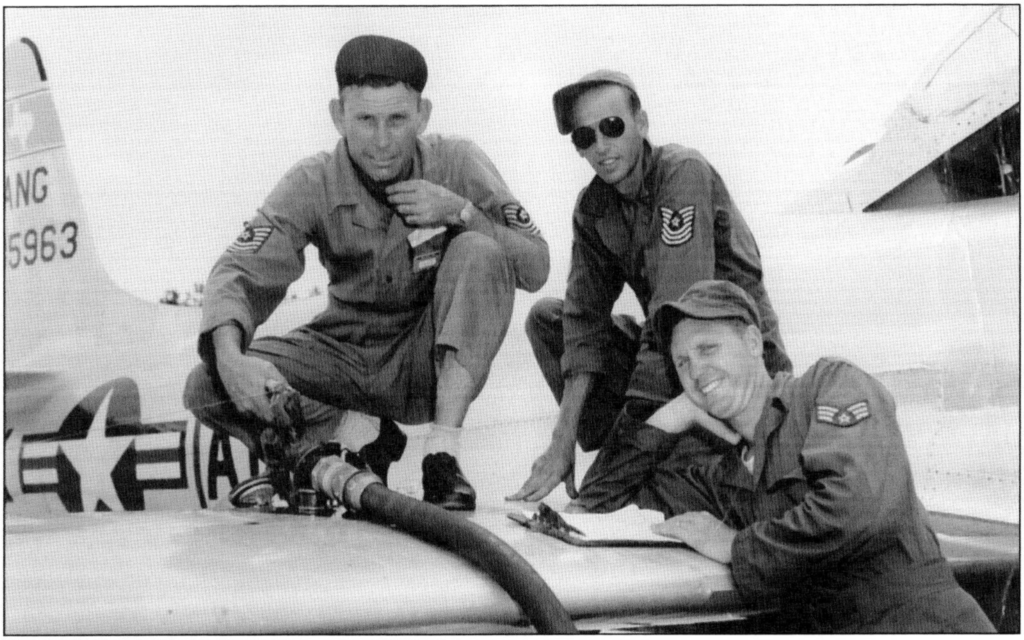

Aircraft had to be fueled to fly. Three mechanics take a break from fueling an F-80 to mug for a photographer. These birds burned JP-4, a 50-50 blend of kerosene and gasoline. (147th.)

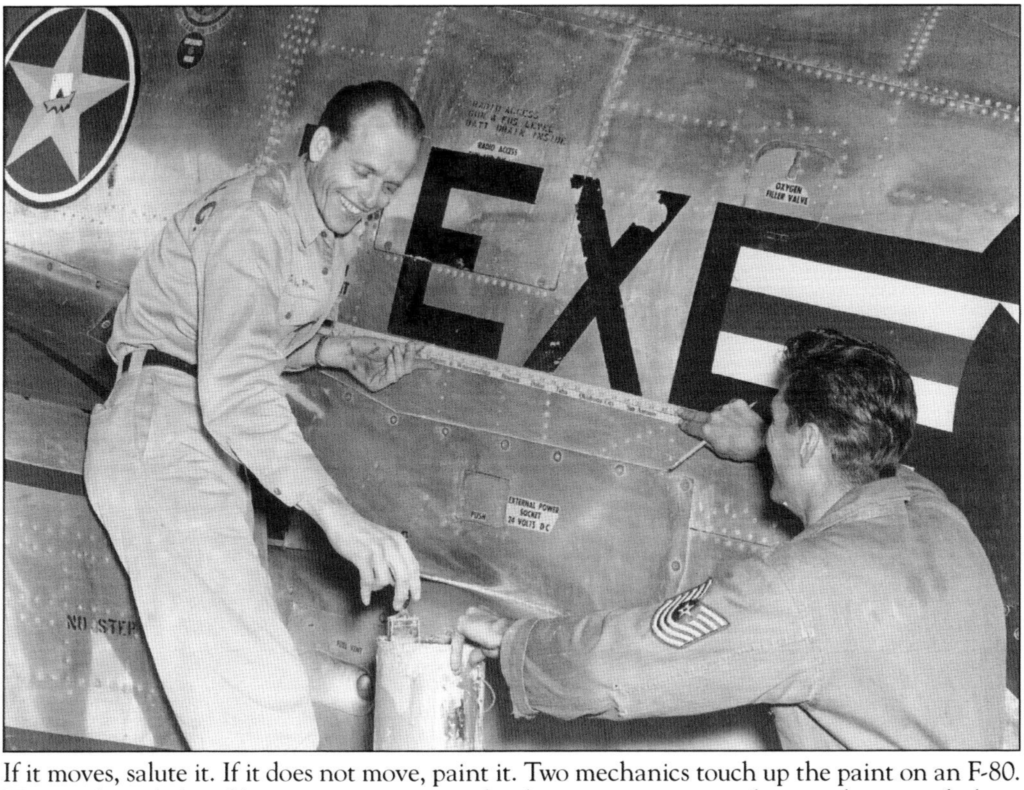

If it moves, salute it. If it does not move, paint it. Two mechanics touch up the paint on an F-80. The 111th prided itself on its appearance, and as long as a jet was standing on the ground, there was an opportunity to improve its looks. (147th.)

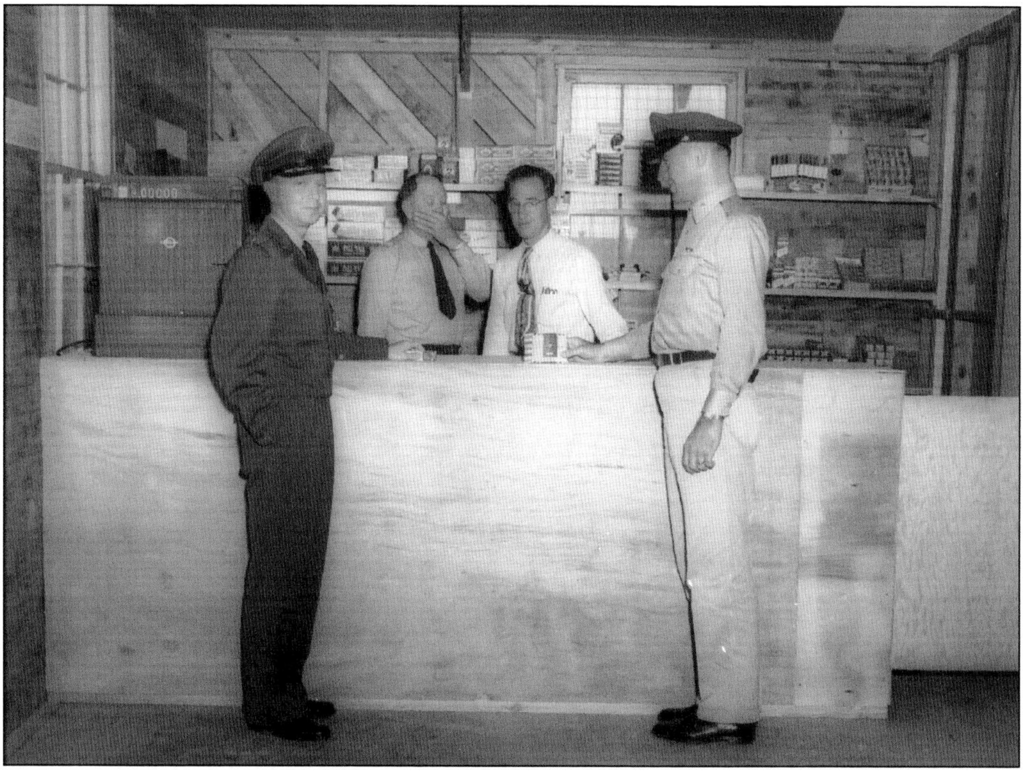

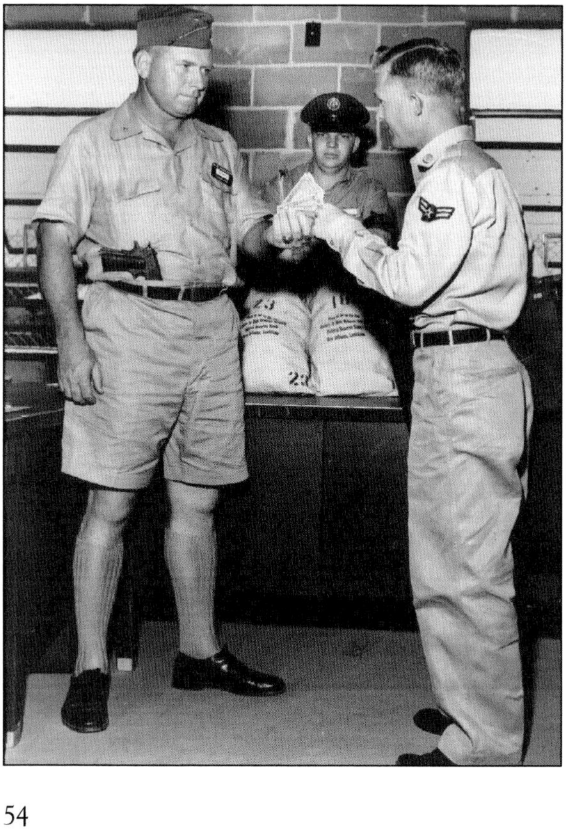

Life as an airman at Ellington was little different than it had been during World War II. The canteen, where airmen could buy smokes and such, looked a lot like the supply rooms from World War II. (LSFM.)

Payday at Ellington is seen here in the 1950s. There was no direct deposit. Ellington's airmen and airwomen received their pay in cash, from the paymaster. He would be armed to prevent any would-be Willie Suttons from making off with the payroll. And those bags were filled with cash. This airman's monthly pay was less than $100 per month. (147th.)

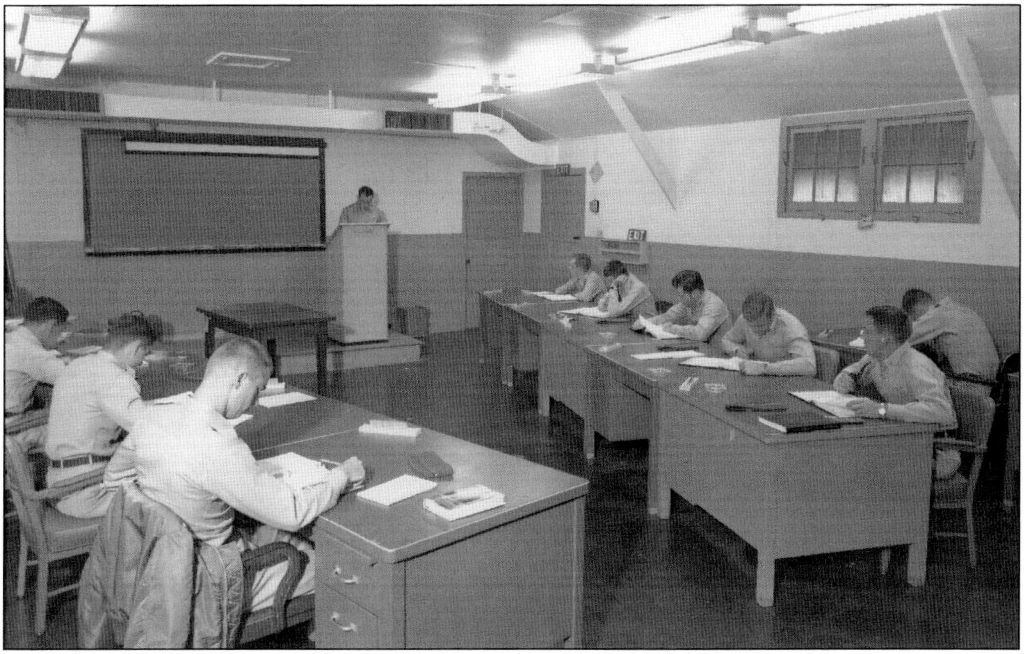

Some things had changed. There tended to be more classroom work in the 1950s than in the 1940s, and students learned at steel desks instead of the wooden tables of World War II. (147th.)

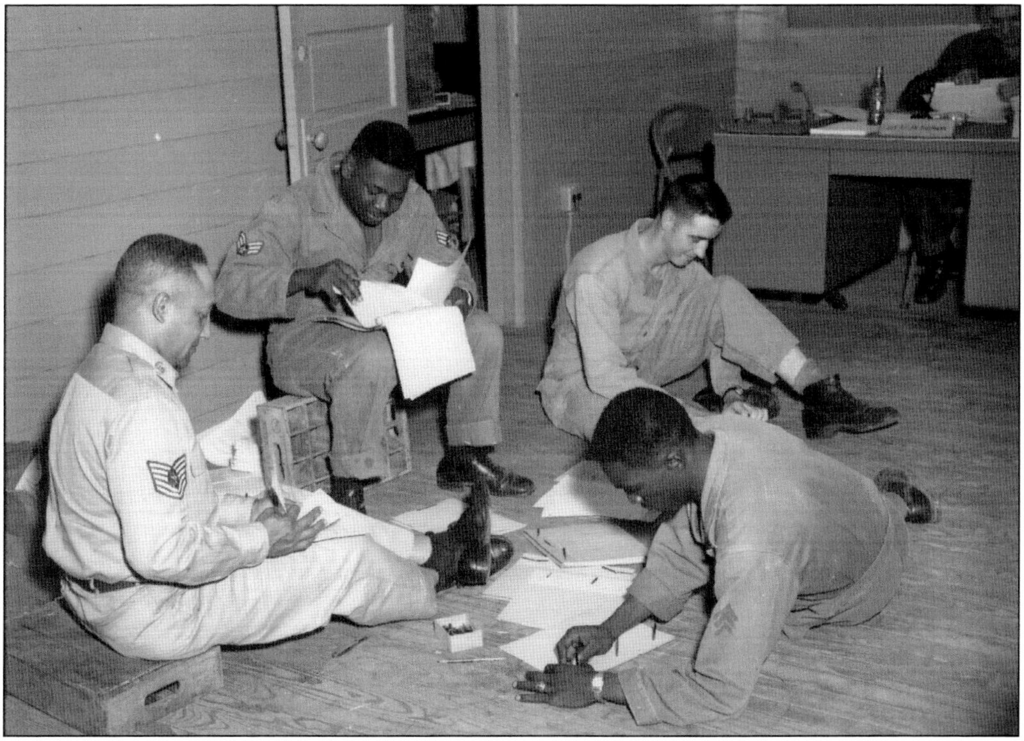

The Air Force was also integrated, and had been since its inception. In the 1950s, black and white members of the Air Force worked and lived together, even in states like Texas where Jim Crow laws were still in force. (LSFM.)

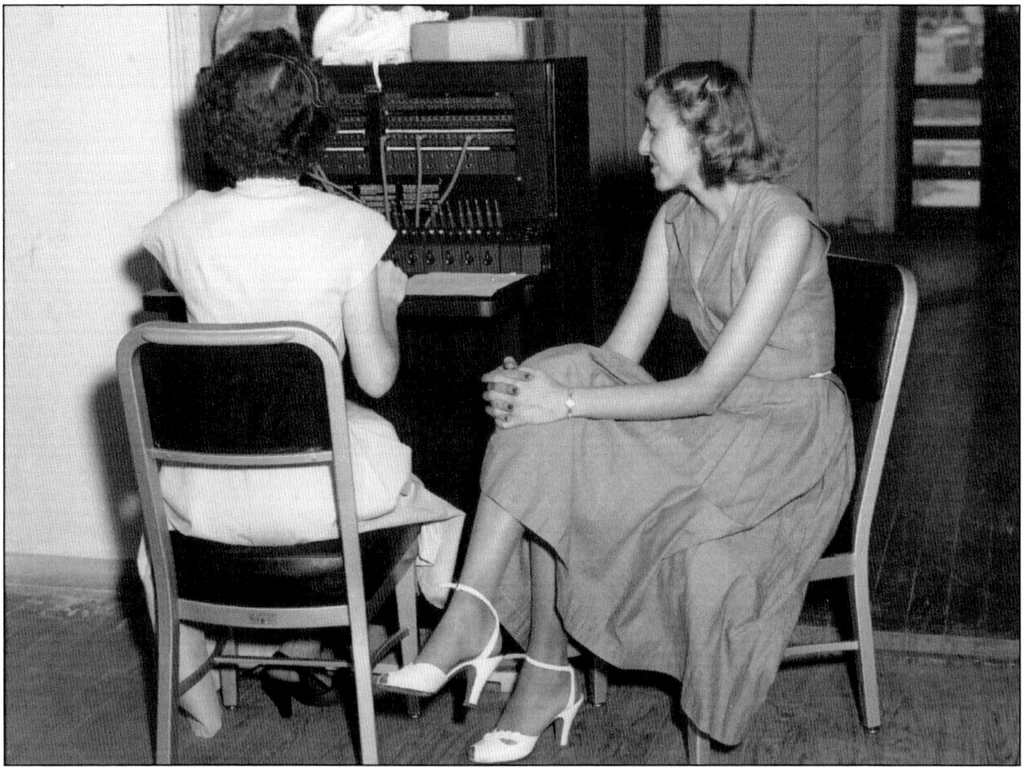

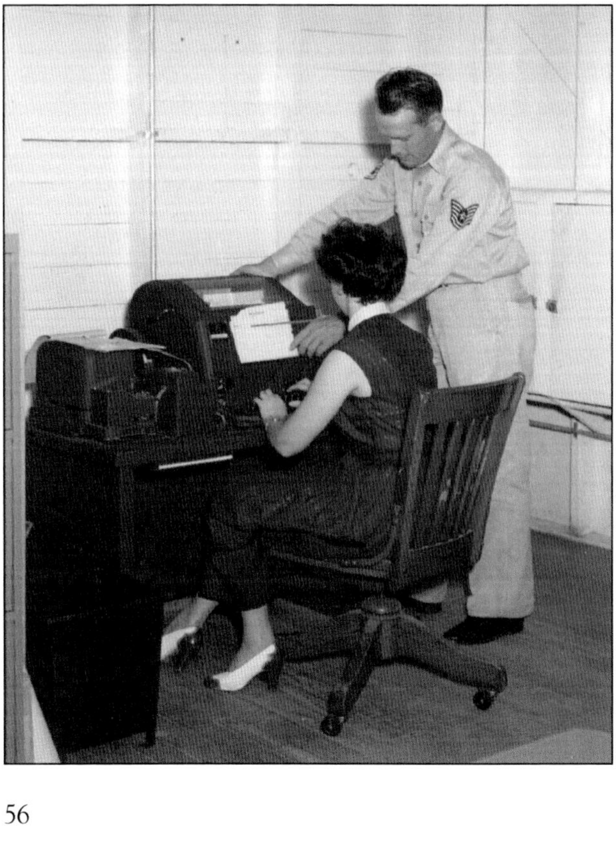

Women continued to work at Ellington, just as they had in World War II. However, most, including this telephone switchboard operator, were civilian contractors rather than WACs. But women in the Air Force were full-fledged members of the Air Force, not an auxiliary corps. (LSFM.)

Automation and electronics were beginning to play a role, although high-tech tended to be teletypes and keypunch machines rather than programmable digital computers. Those would not arrive until the 1960s and 1970s. (LSFM.)

It was a glorious thing to be a fighter pilot, whether in the 1950s or the 21st century. This lieutenant poses next to his jet with a look intended to let you know he was the hottest man in the skies. (147th.)

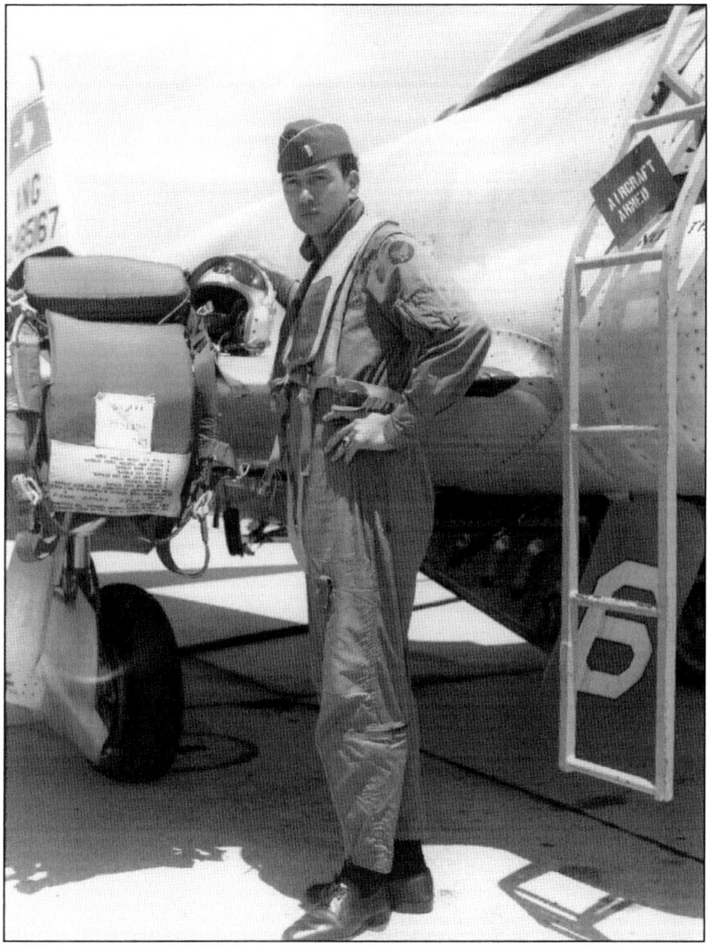

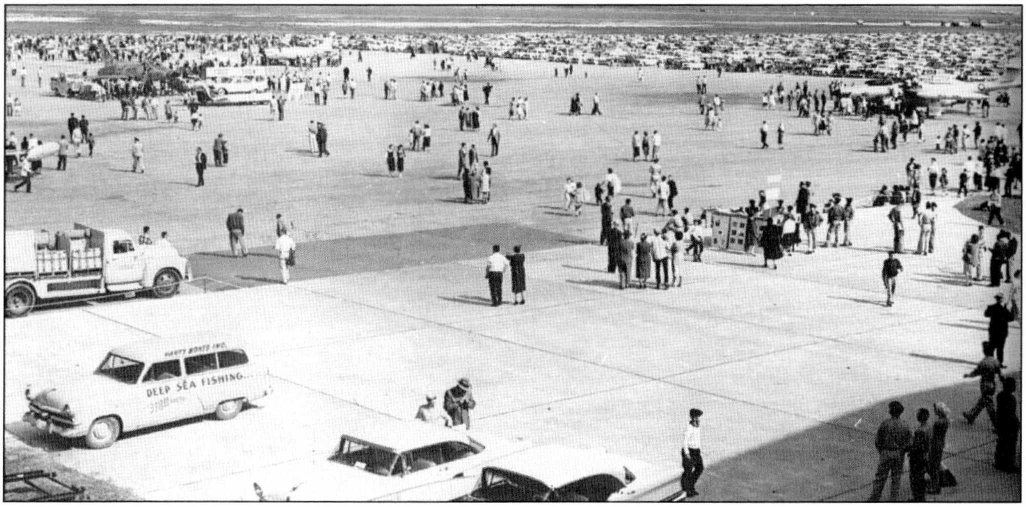

Ellington Air Force Base, with its large Air National Guard contingent, held occasional open house days to give family and the local community an opportunity to visit the base and see the Air Guard at work. This photograph was taken during a 1958 open house. (Missy Bryant.)

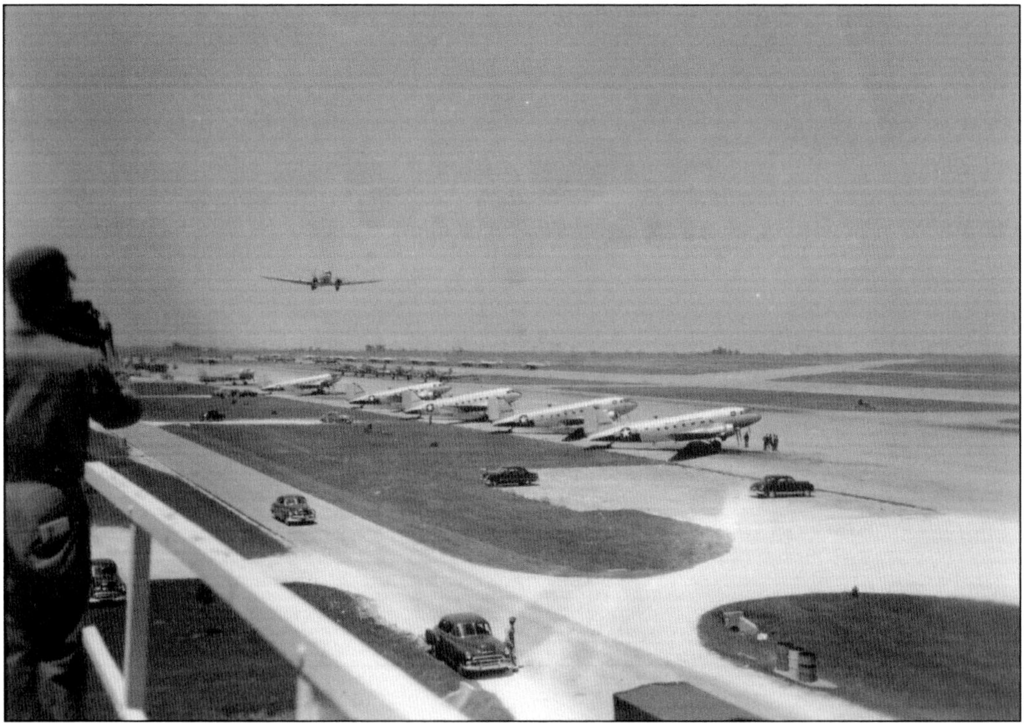

Ellington Air Force Base is seen here in the late 1950s or early 1960s. An Air Force C-47 shoots a landing on runway 5-17, while others are parked on the apron during an exercise. (LSFM.)

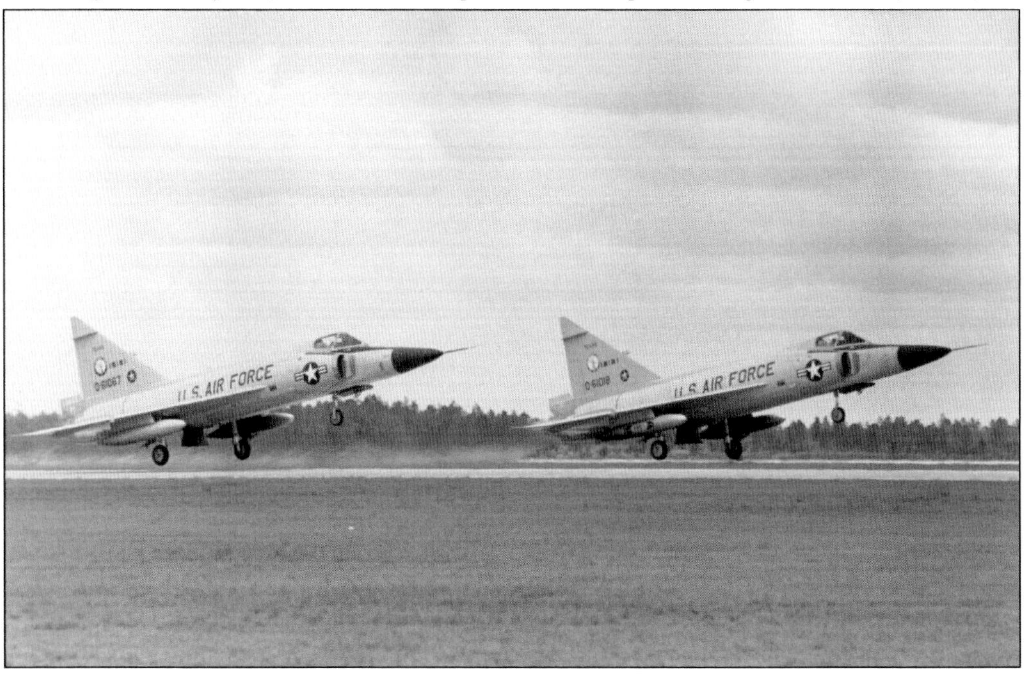

The F-102 Delta Dart replaced the F-86 in 1960. This was an all-weather supersonic interceptor, one of the hottest aircraft of its day. Ellington would continue flying the F-102 until 1974. Here, two 111th Squadron F-102s take off from Ellington to start a patrol. (NMUSAF.)

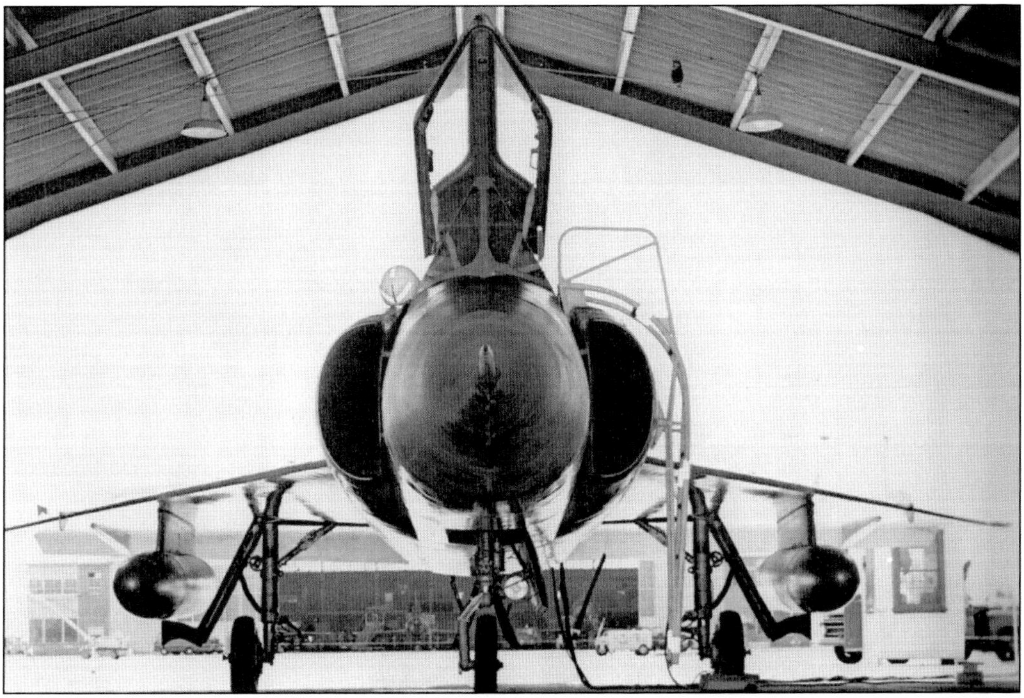

An F-102 is pictured on alert on the ground at Ellington. Since no one knew when World War III might start, the Air Force made it a policy to have aircraft ready to scramble at any time. Once notified, a pilot would board the fighter, close the cockpit, and take off, all within minutes. Ellington maintained this posture from 1958 through 1974. (LSFM.)

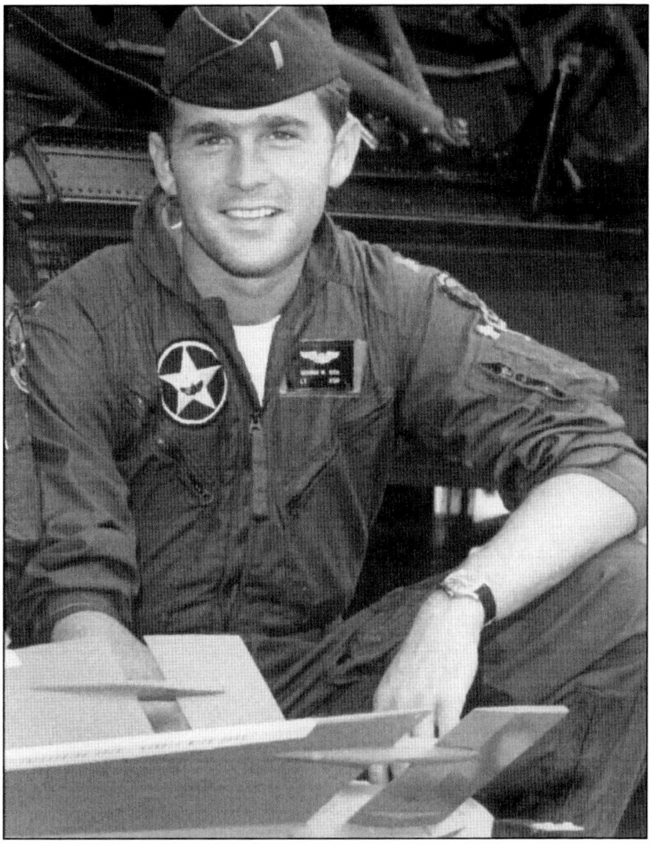

One of the pilots flying for the 111th Squadron was this man, Lt. George W. Bush. He flew F-102s from 1970 through 1972. Afterwards, he went on to a career as a West Texas oil man, owned the Texas Rangers baseball team, entered politics, and became the 43rd president of the United States. (DoD.)

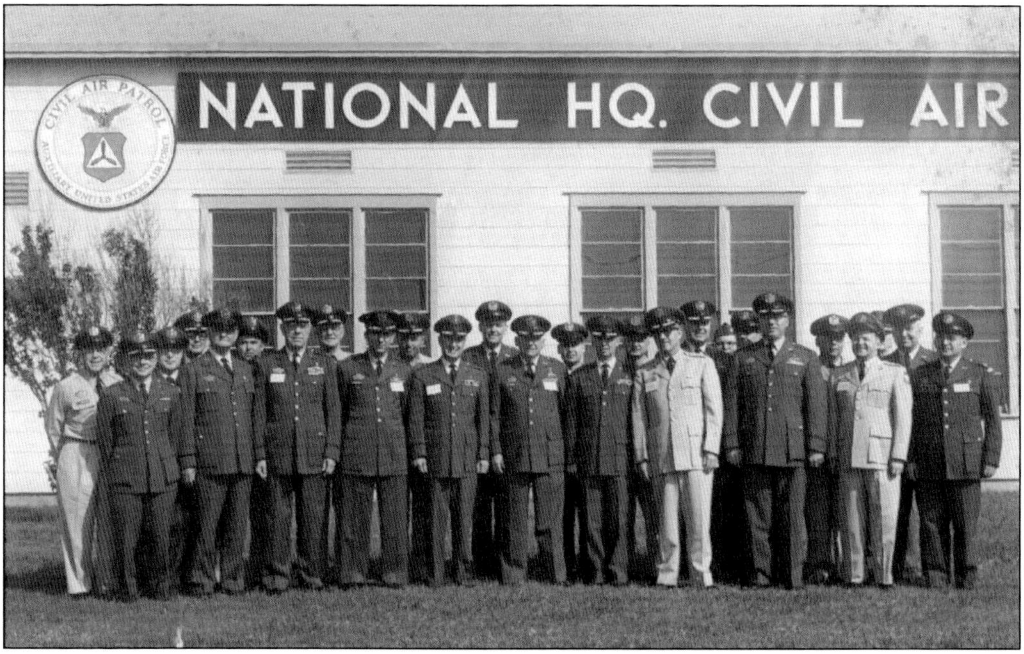

The national headquarters of the Civil Air Patrol, the Air Force civilian auxiliary, was at Ellington from 1959 through 1967. It was housed in this building. CAP headquarters moved to Maxwell Air Force Base in 1967 because the Ellington headquarters were not air-conditioned. CAP had leased an IBM computer, and IBM required it be housed in an air-conditioned building. (LSFM.)

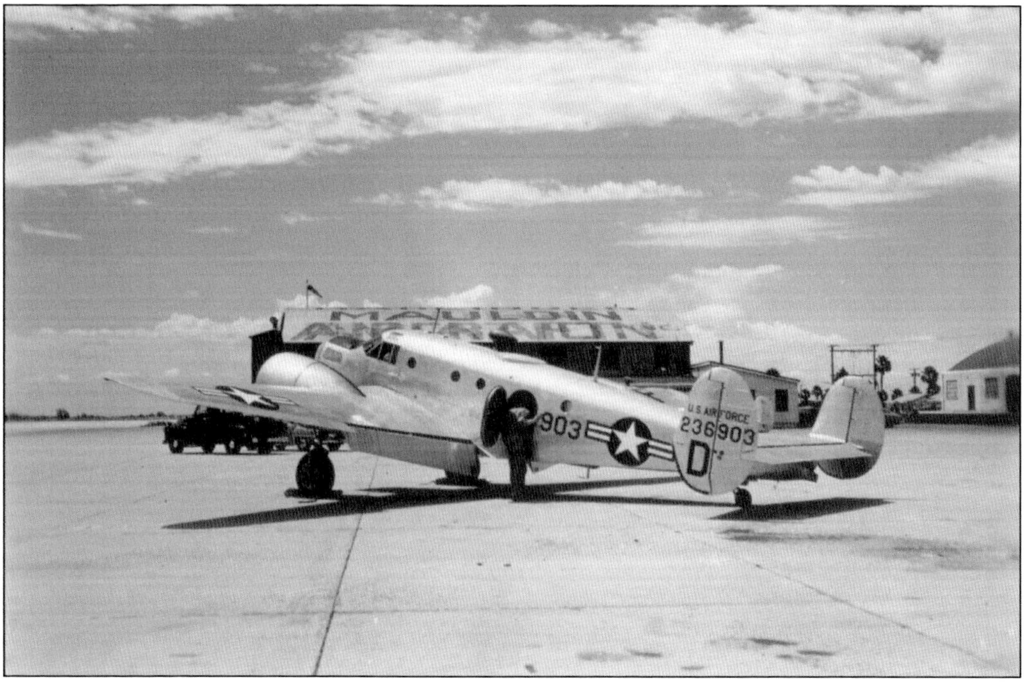

A visiting "fireman" (important Air Force official) arrives at Ellington in an executive transport. Ellington served as a convenient stopping point for any official in the armed services flying to Houston. (LSFM.)

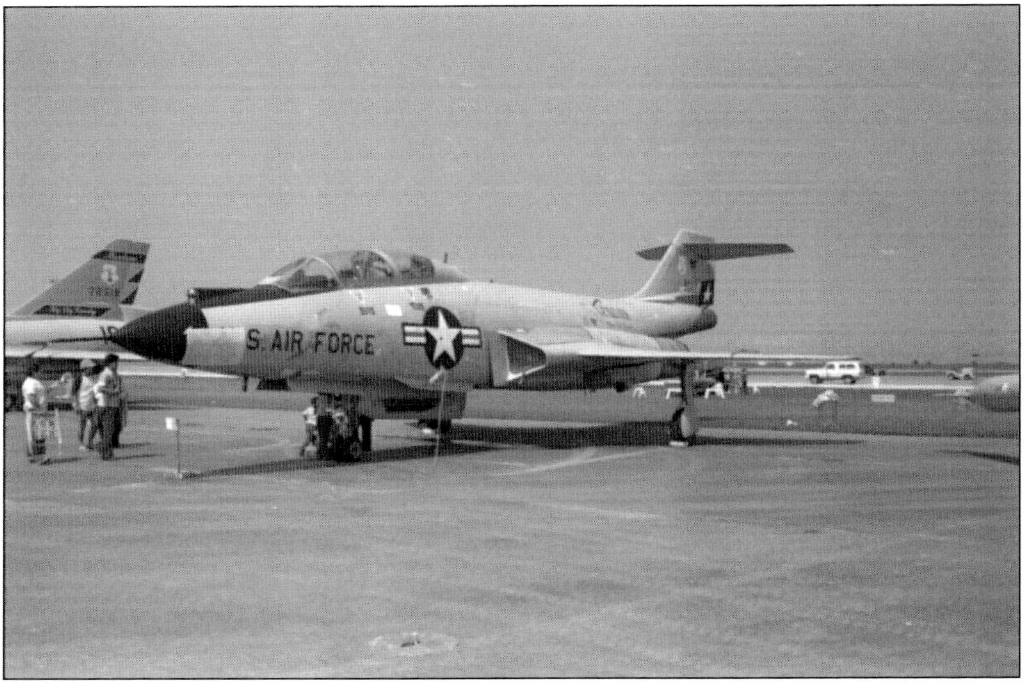

In 1971, the 111th began operating the F-101 Voodoo, another supersonic Century Series fighter. Ellington had both the single-seat F-101B fighter version and the tandem seat F-101F (pictured), which was also used for reconnaissance. Ellington operated F-101s until 1982. (Shawn Harrison.)

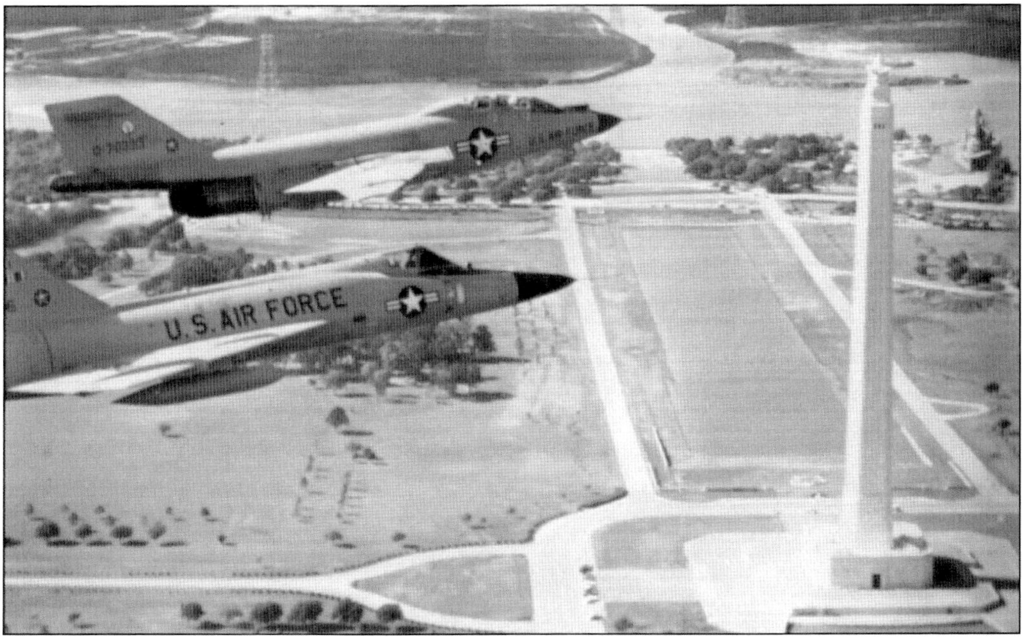

For a few years, the 111th was flying both the F-101 and the F-102. This photograph shows the two fighters flying past the San Jacinto monument at San Jacinto Battlefield Park. (LSFM.)

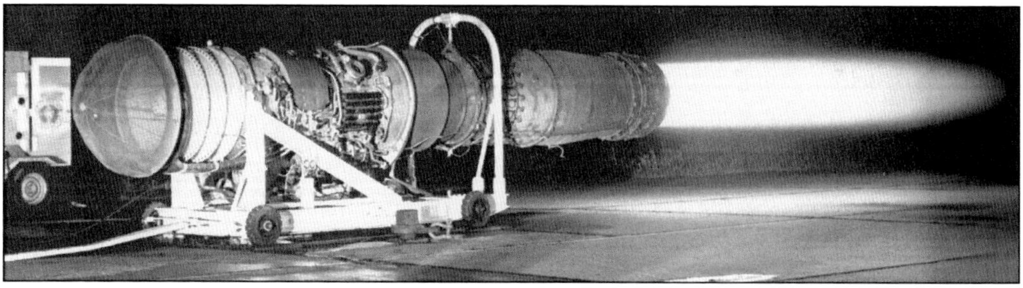

As with its earlier aircraft, Ellington had all the facilities required to maintain its aircraft in flying condition. Here, an engine from one of its Century Series fighters is checked out during a test in Ellington's maintenance hangar. (LSFM.)

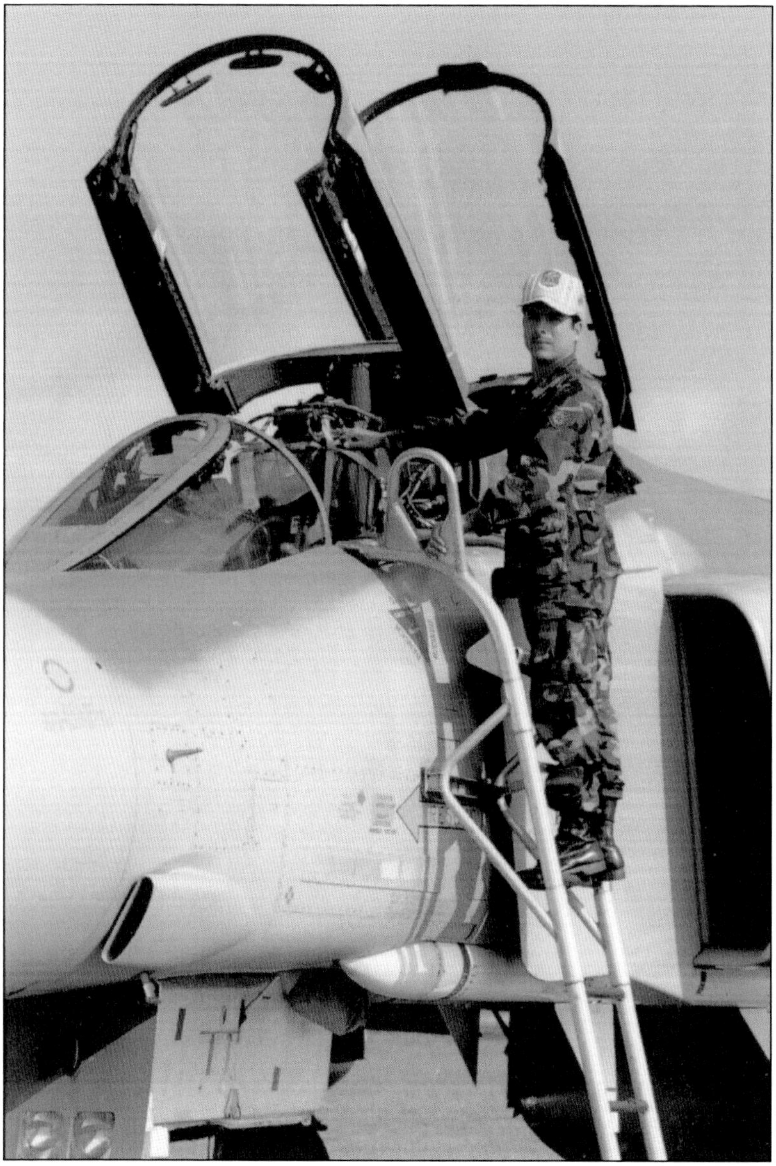

In 1982, the F-101 Voodoos were replaced with F-4C Phantoms, a Vietnam-era fighter-bomber. Ellington would operate Phantoms until 1989, when they were replaced by the F-16 Fighting Falcon. (147th.)

Four

NASA's Airfield

1962–Present

The National Aeronautics and Space Administration presence at Ellington is almost as long as NASA's existence. NASA was founded in 1958, shortly after the Soviet Union ignited the Space Race with their October 1957 launch of Sputnik I, the world's first artificial satellite. One of NASA's first missions was to put a man in space. It established a Manned Spaceflight Center temporarily at its center in Langley, Virginia, planning to build a permanent facility somewhere else.

In 1961, NASA named Houston as the permanent site of the Manned Spacecraft Center. One reason was Ellington Air Force Base. One criterion was the presence of a military airfield to support NASA activities. But in 1961, the planned site for the Manned Spaceflight Center, today's Johnson Space Center, was a cow pasture. The campus had to be built. Starting in early 1962, while the MSC was under construction, Ellington was used as the MSC's airfield and offered a place to house laboratories and offices that had to be on a federal facility.

Even after the MSC was completed and it became JSC, Ellington housed NASA's fleet of aircraft based in Houston and maintenance facilities for them, and provided space for laboratories and offices for NASA's Houston-area aeronautics-related activities, generally atmospheric research.

NASA moved most of its space-related training facilities, offices, and laboratories out of Ellington in the 1960s and 1970s as the buildings to house them at the Space Center were completed. Eventually, Ellington's main NASA function was to serve as JSC's airbase. In the 1990s, NASA established its neutral buoyancy training facility on Ellington's undeveloped south side. It was almost accidental. NASA purchased a building originally intended as a space station integration and assembly facility. The building was repurposed for extravehicular activity training.

Today, NASA and Ellington continue their 55-year partnership. Astronauts train there, both to maintain atmospheric flight proficiency and for space missions. NASA aircraft continue to fly from Ellington, and plans are now afoot to use Ellington as a launch site for spacecraft, both NASA and commercial.

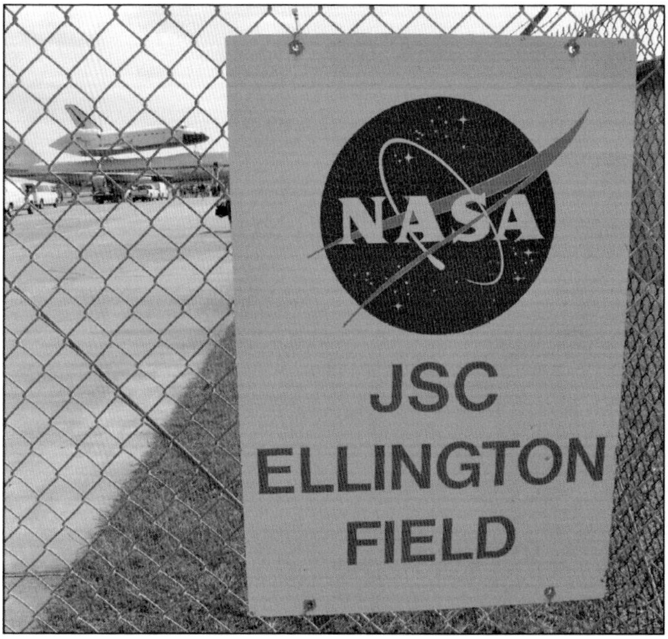

NASA has been at Ellington Field for so long that when someone mentions Ellington, most people think of NASA. Only those at or older than retirement age can remember a time when NASA lacked a presence here. (DVIDS.)

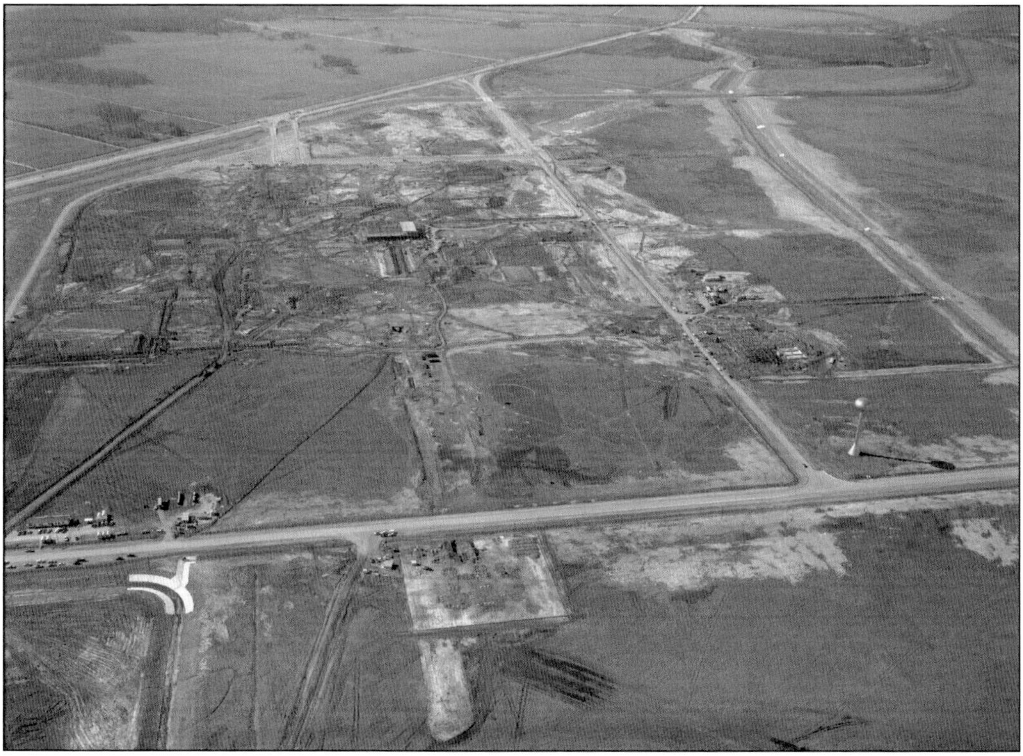

NASA's presence at Ellington dates to 1962. The agency moved the Manned Spaceflight Center to Houston in 1961. There were no facilities at the site of what is now the Johnson Space Center. It was a cow pasture. This photograph, taken in 1962, records the center's appearance when construction began. In addition to serving as NASA's airport, Ellington provided facilities for laboratories that could not be placed outside a federal facility. (NASA.)

Today, in the 21st century, nearly 60 years after NASA's arrival in Houston, Ellington remains an integral part of JSC, both serving as JSC's airport and as a location for NASA research activities. (DVIDS.)

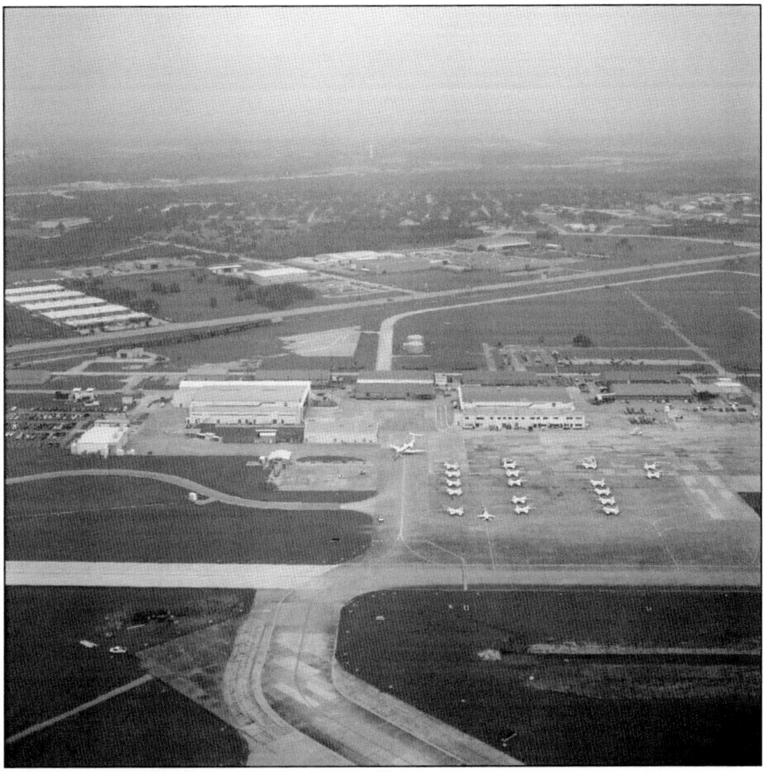

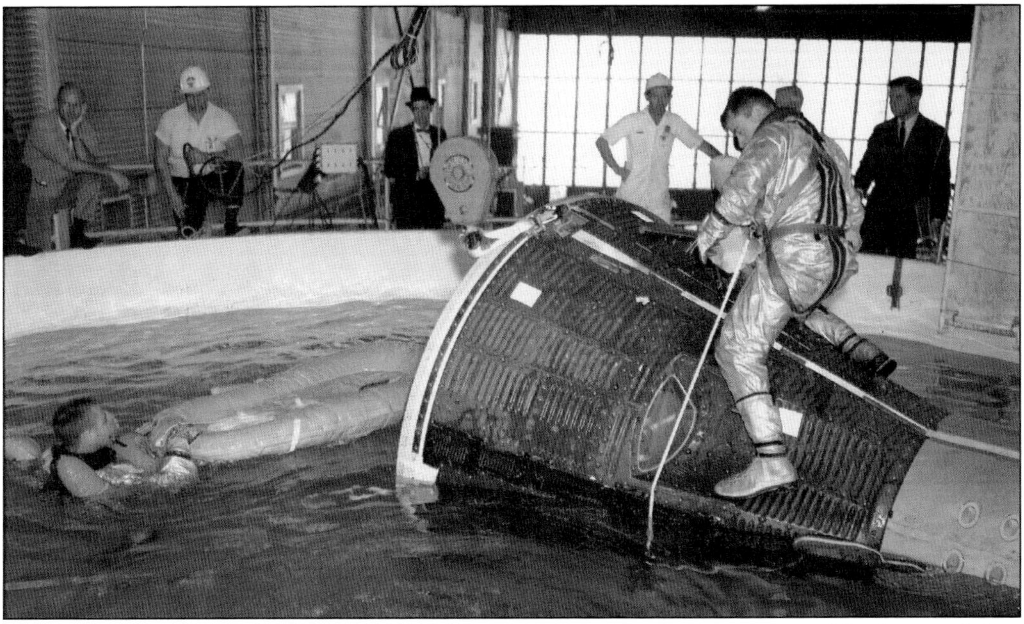

One of NASA's earliest uses for Ellington Air Force Base was water training. Not for weightless environment training, but simpler water egress training. Mercury, Gemini, and Apollo capsules landed in the ocean. After one Mercury capsule sank, almost drowning an astronaut, NASA expanded training on leaving the capsule. This is John Young practicing prior to the Gemini 3 mission. (NASA.)

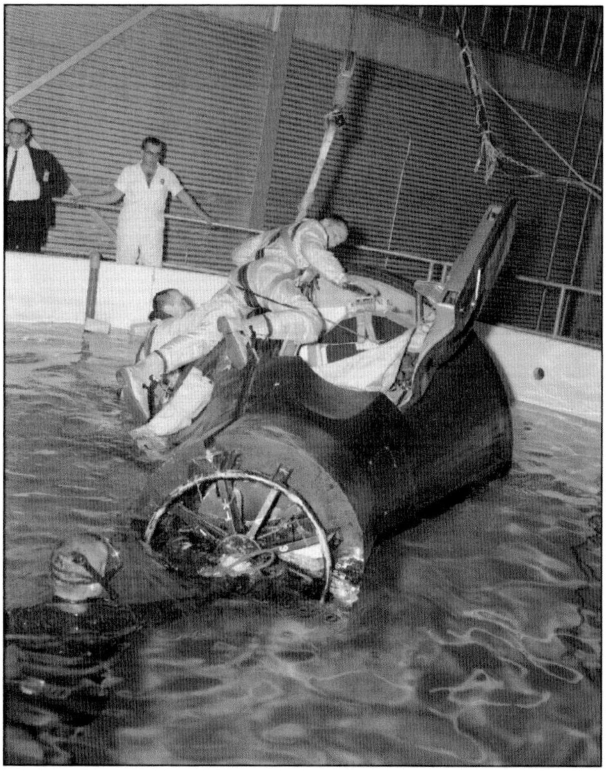

Water egress training started during the Gemini program. A special trainer was built, which is shown here. The trainer, known as Gemini Static Article No. 5 (SA-5) was first used in September 1964. This picture shows it being used in 1965 by the Gemini 4 prime crew, James McDivitt and Ed White. (NASA.)

Ellington was also used for water egress training during the Apollo program. An Apollo capsule trainer similar to the SA-5 was built and used at Ellington. Pictured here are the Apollo 1 crew, Gus Grissom (left), Ed White (foreground, right) and Roger Chaffee (behind White) in rafts during a June 1966 session. An outdoor swimming pool at Ellington was used. (NASA.)

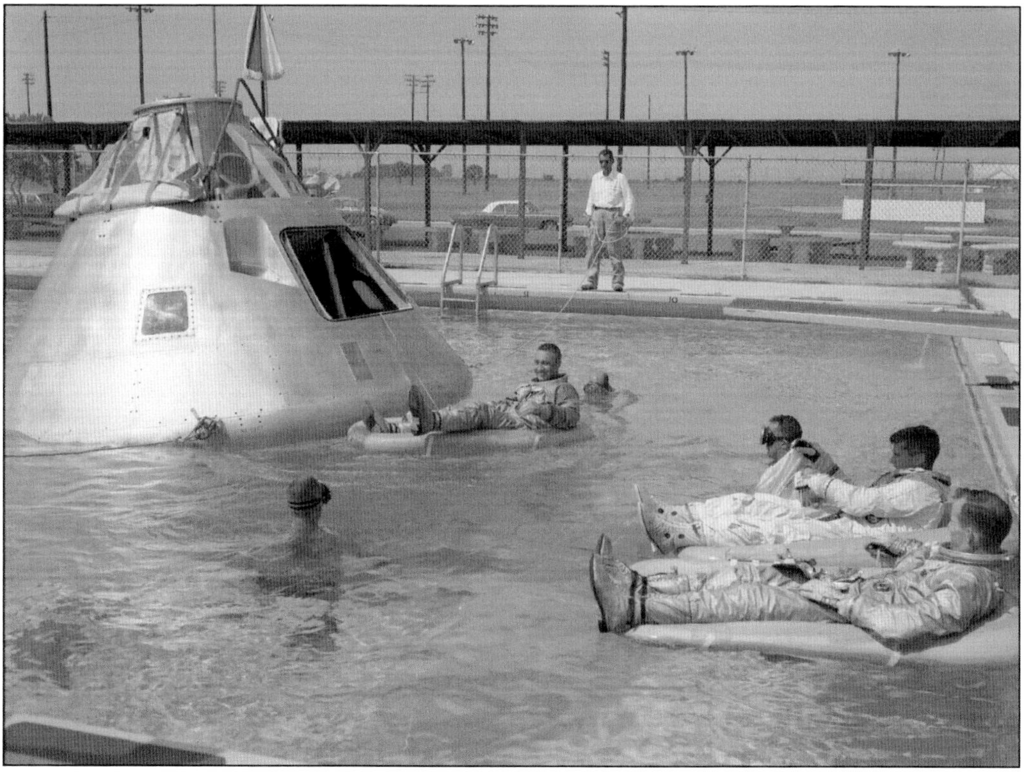

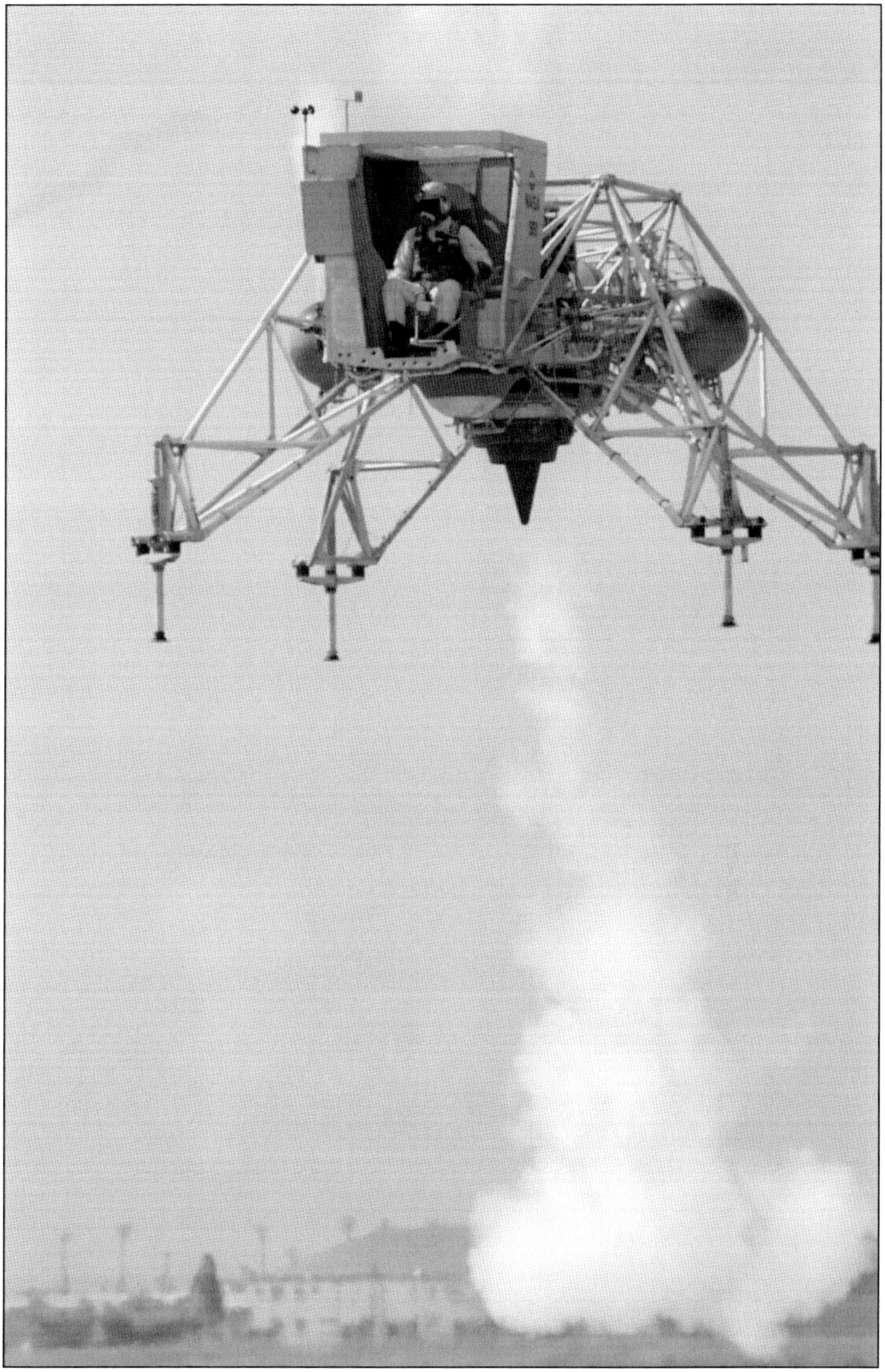

Ellington was also where Apollo astronauts practiced moon landings. Here, Charles Conrad, commander of the Apollo 12 mission, is flying the Lunar Landing Training Vehicle during a 1969 training session. The LLTV was designed to simulate flying in the moon's one-sixth gravity, airless environment. It was highly unstable in Earth's gravity. (NASA.)

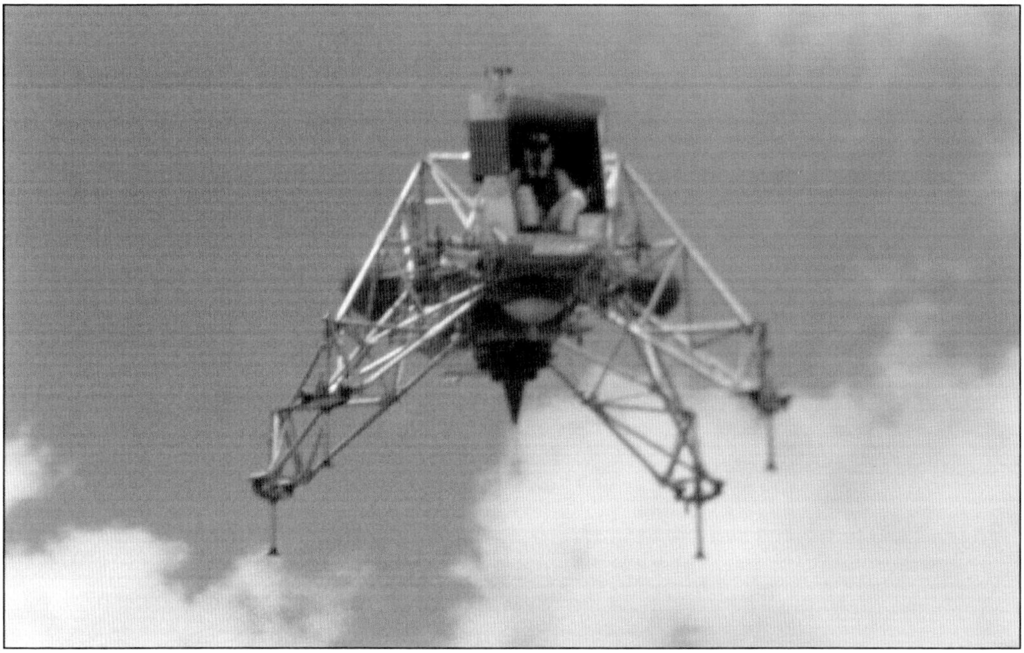

How dangerous the LLTV could be was demonstrated on May 6, 1968. That day, Neil Armstrong made his 22nd flight on the simulator. This image was transferred from a 16mm film of the flight, hence the low quality. This frame was captured in the first five minutes Armstrong was airborne. (DVIDS.)

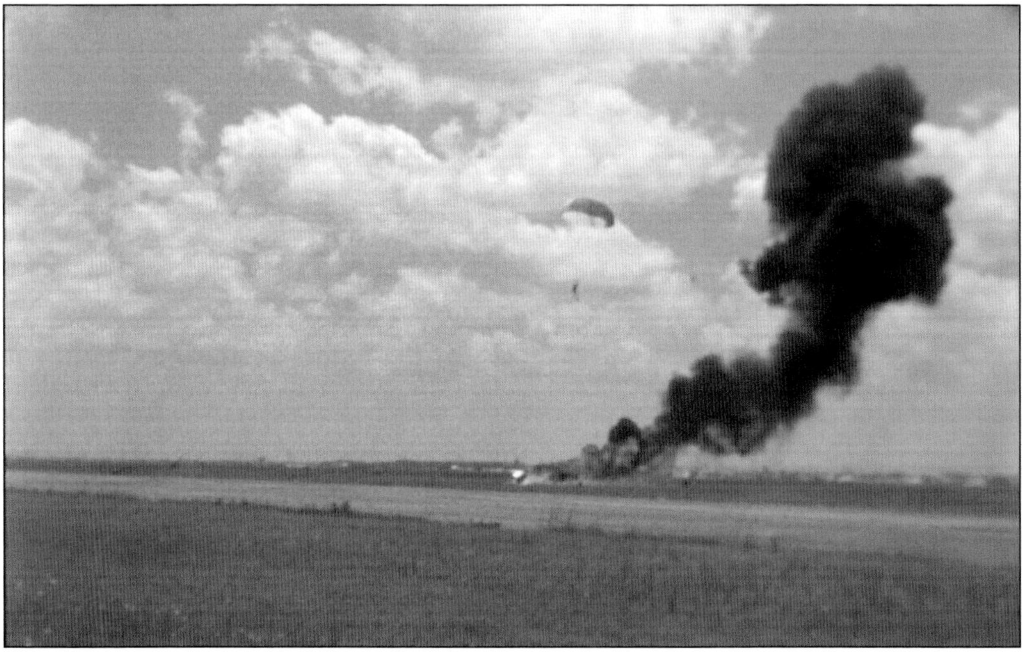

Just before the five-minute mark, the control system failed. The vehicle tumbled to the ground, exploded, and burned. Armstrong ejected seconds before it crashed, only 200 feet above the ground. His parachute can be seen above the flaming wreckage. His cool behavior was one reason Armstrong was picked to command the first lunar landing mission. (DVIDS.)

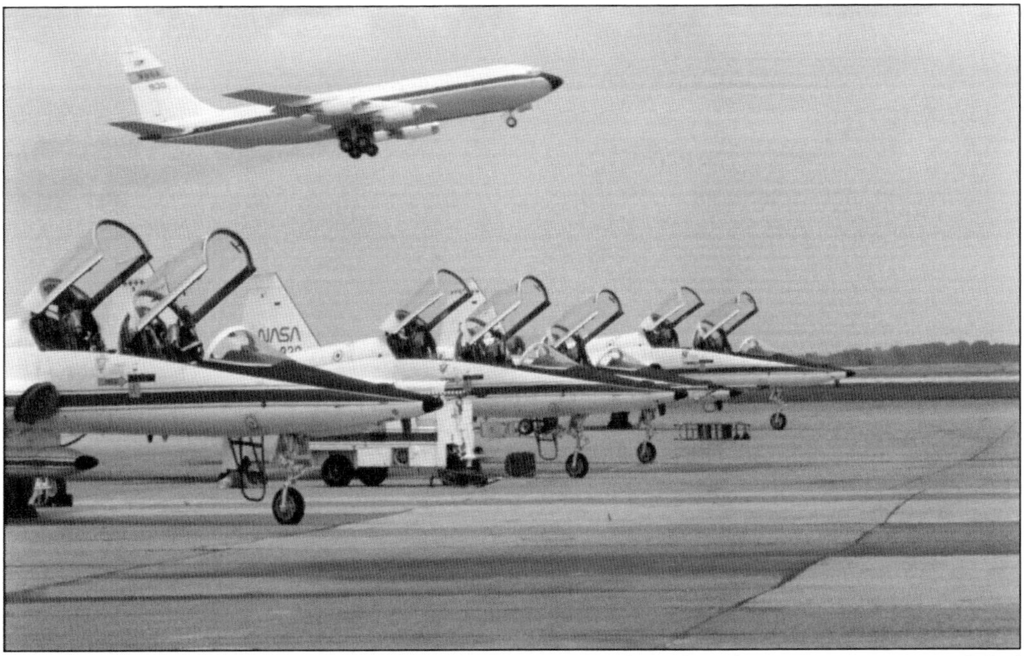

Another crew training tool operating out of Ellington was the reduced-gravity trainer, a modified KC-135, known among those who flew on it as the "Vomit Comet." It is shown taking off from Ellington in the 1980s. It would fly over the Gulf of Mexico, where it would fly parabolas over the water. For about 90 seconds, passengers would experience weightlessness. (NASA.)

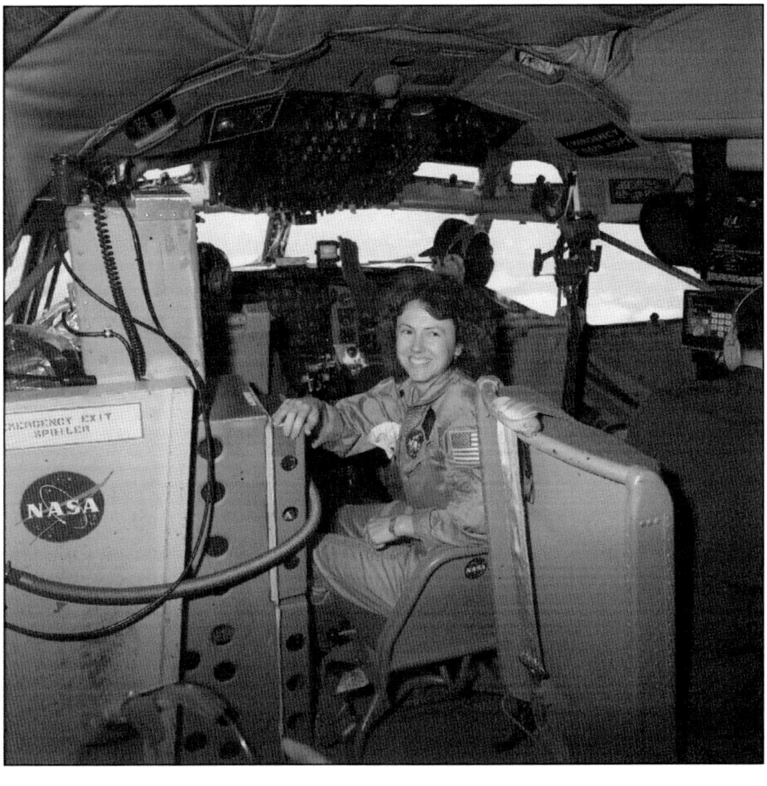

All astronauts, including space shuttle mission specialists and payload specialists, trained on the KC-135 to experience weightlessness. Each training flight went through 40 to 60 parabolas. This photograph shows Christa McAuliffe, selected for the first teacher-in-space shuttle mission, in the reduced-gravity trainer as part of her preparation for the ill-fated 51-L mission. (NASA.)

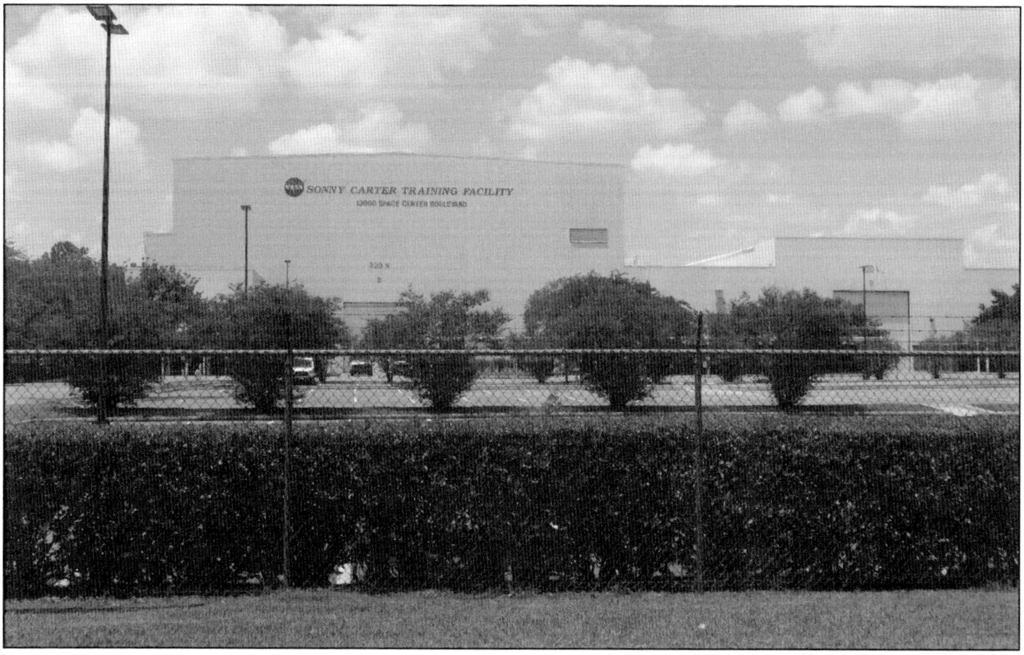

The Sonny Carter Training Facility is also located at Ellington, albeit on the south side off Space Center Boulevard. It was built by McDonnell Douglas as a facility to assemble Space Station Freedom components, but stood empty after Freedom was cancelled. NASA purchased the building in 1995. (AP.)

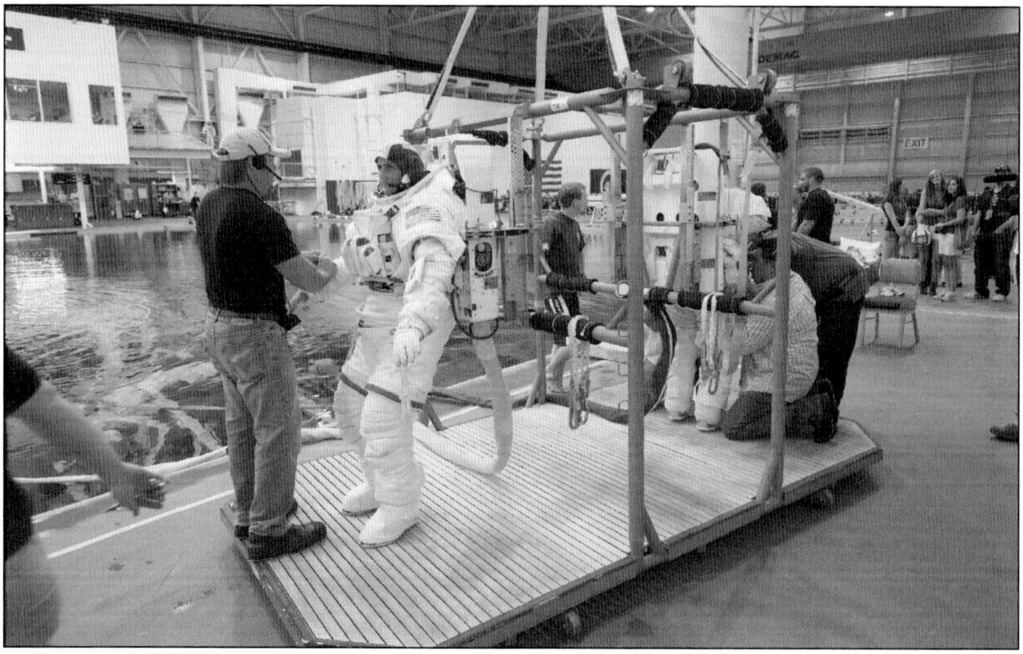

NASA installed a massive, 202-foot by 102-foot pool that is 40 feet deep. It is now the Neutral Buoyancy laboratory, used to train astronauts how to conduct extravehicular activities for upcoming space missions. Here, two astronauts go through a briefing before entering the pool to practice space station assembly activities. (NASA.)

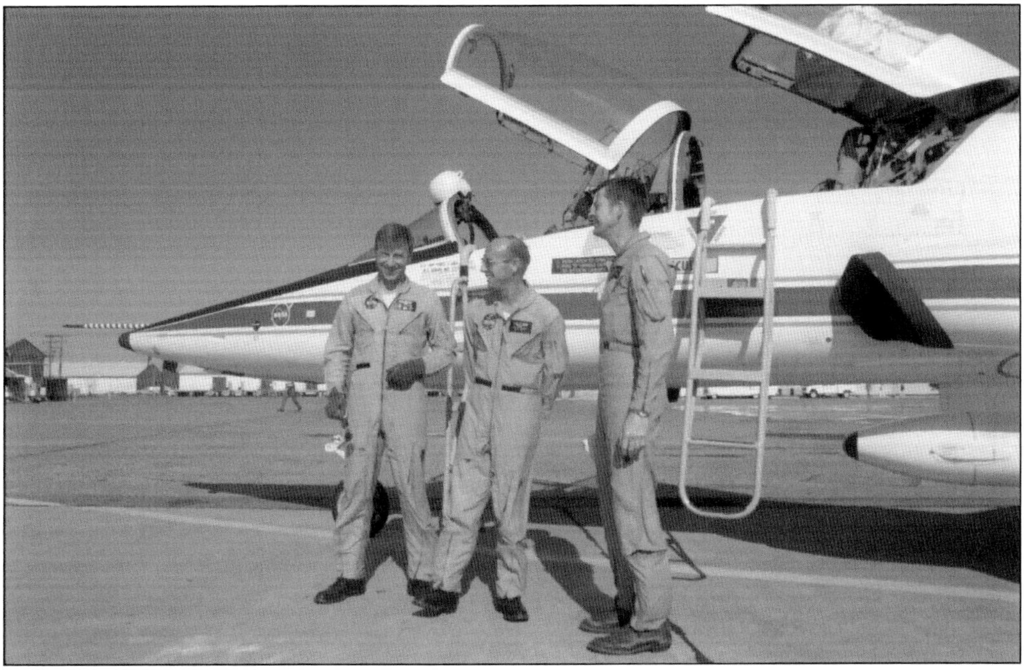

Ellington is also the point of departure for JSC-based astronauts when they leave Houston to head for Cape Canaveral and launch from Kennedy Space Center. Seen here from left to right are Paul Weitz, Charles Conrad, and Joseph Kerwin, crew of the first manned Skylab mission (Skylab 2) standing beside a NASA T-38 jet before departing to KSC before their 1973 launch. (NASA.)

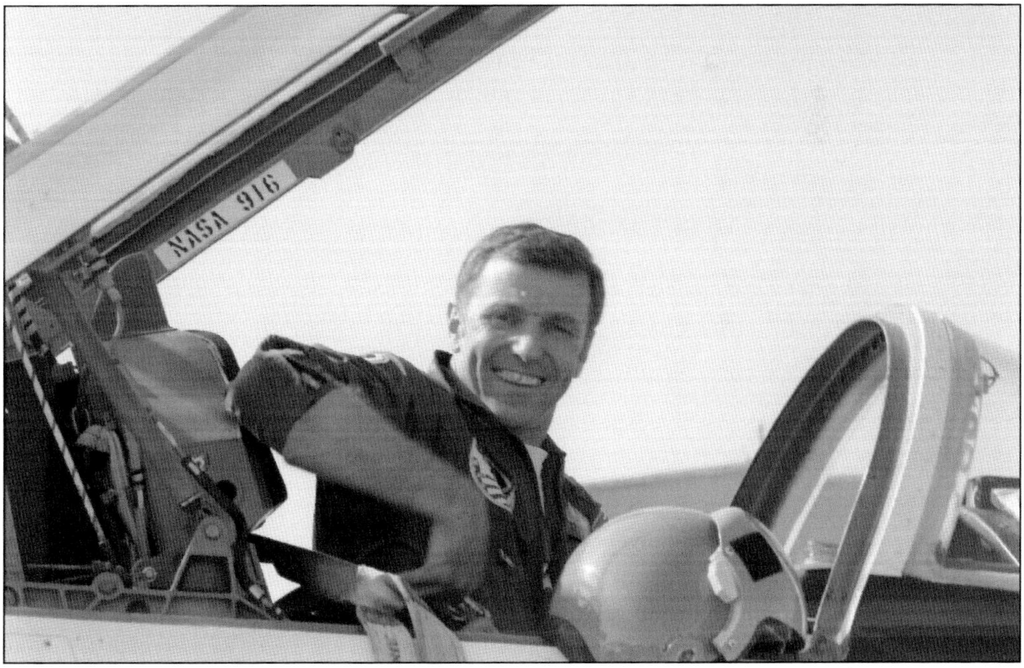

Eight years later, it was the turn of the shuttle astronauts to depart Ellington for KSC. Here is Dick Truly in the front-seat cockpit of a T-38 just before flying to Florida for the STS-2 mission in November 1981. (NASA.)

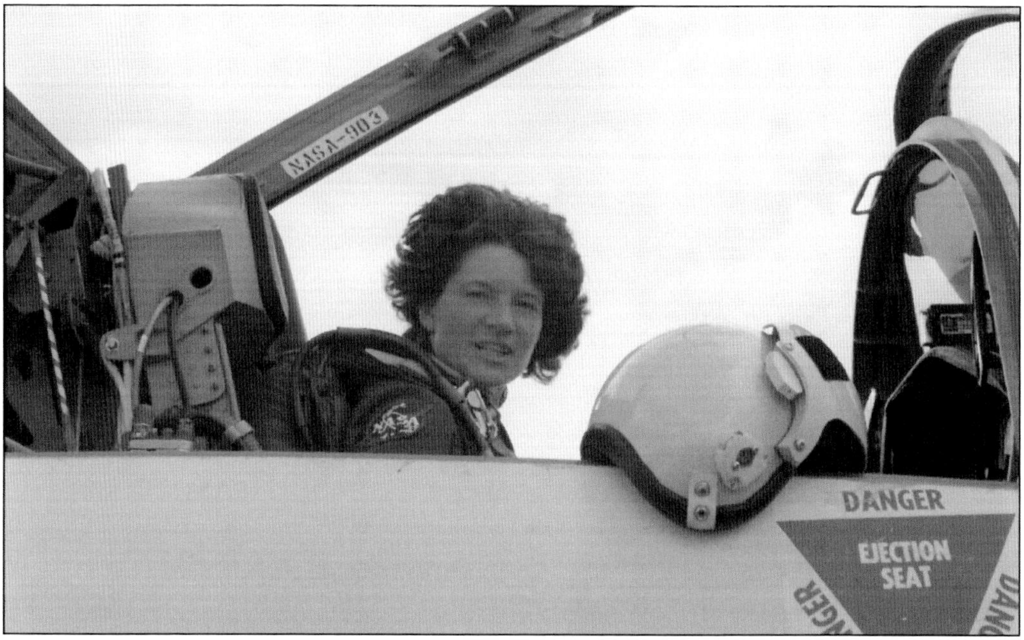

Sally Ride was the first female American astronaut to enter space. A mission specialist, she is sitting in the aft cockpit of a T-38 Talon before departing Ellington to board the orbiter *Challenger* for the STS-7 mission, which launched on June 18, 1983. (NASA.)

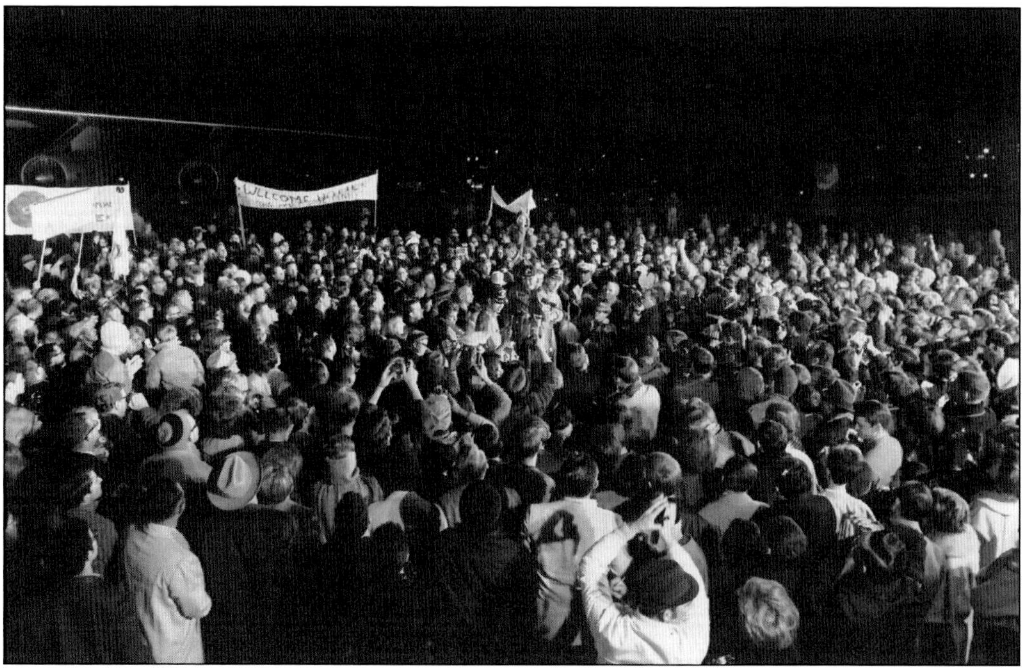

Ellington was also the site of many noted arrivals. The Apollo 8 crew, the first men to orbit the Moon, arrived home to Ellington Air Force Base late at night on December 29, 1968, after the conclusion of their historic flight. Despite the late hour, Frank Borman, James Lovell, and Bill Anders were greeted by a crowd of over 2,000 NASA employees, contractors, and their families. (NASA.)

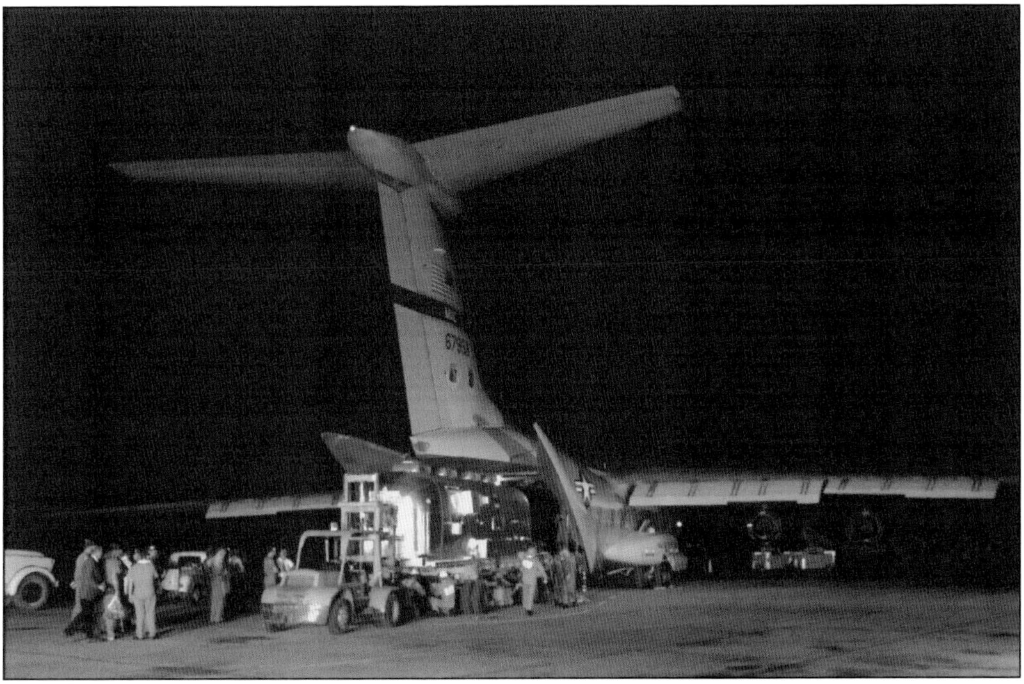

Seven months later, it was the turn of the Apollo 11 astronauts. They, too, arrived at Ellington after dark. The Mobile Quarantine Facility with the Apollo 11 crew inside is being unloaded from an Air Force C-141 transport in the predawn hours of Sunday, July 27, 1969. (NASA.)

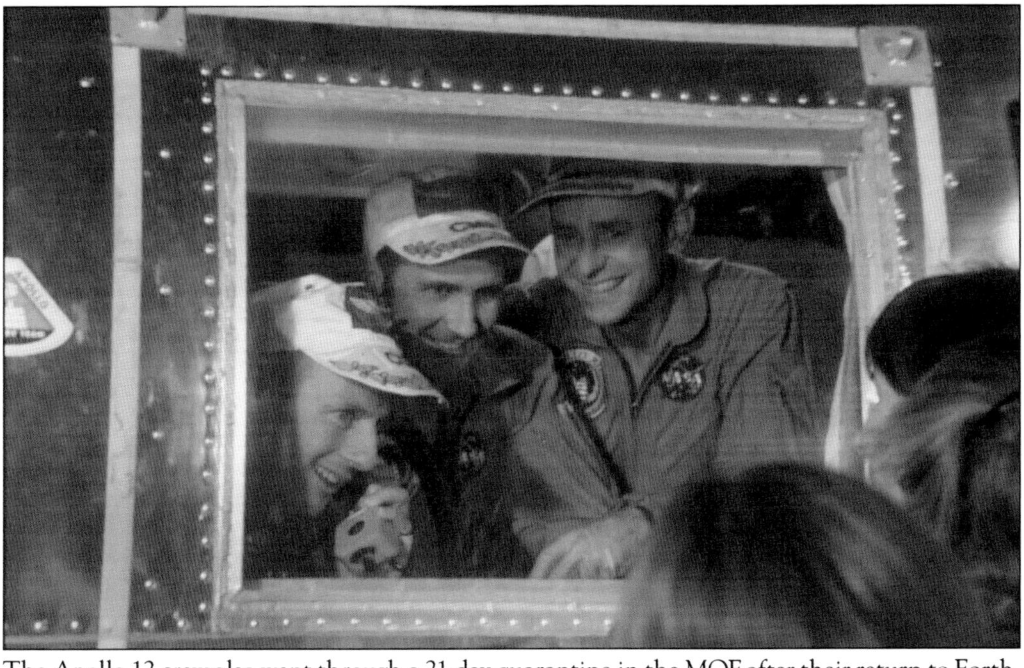

The Apollo 12 crew also went through a 21-day quarantine in the MQF after their return to Earth. This picture shows the crew in the facility, a modified Airstream trailer, greeting their families at Ellington. They are speaking by telephone (the orange item held by astronaut Charles Conrad). The trailer arrived at Ellington November 29, 1969. (NASA.)

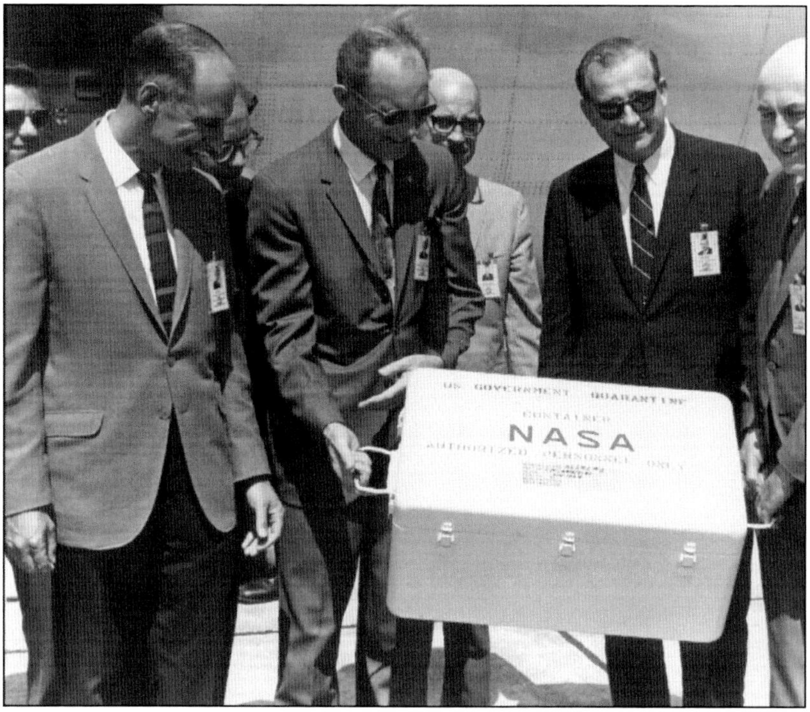

Ellington was also where the Apollo moon samples arrived in Houston. JSC officials pose for photographers with the first container of Apollo 11 lunar samples to arrive at Ellington before taking the box to the Lunar Sample Processing Facility at JSC. (NASA.)

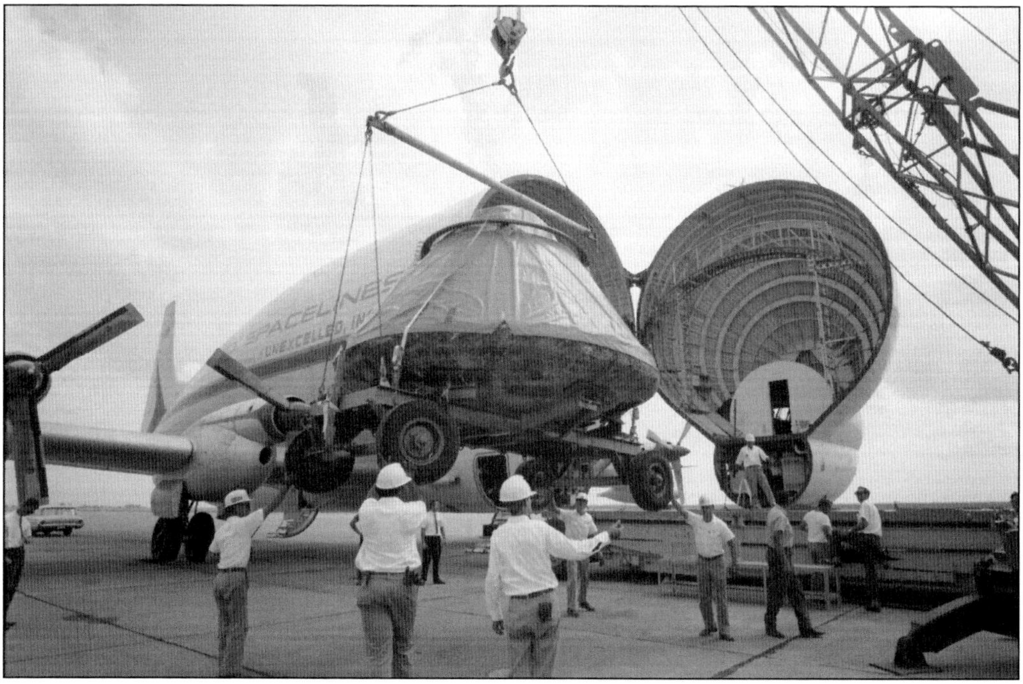

The Apollo 11 capsule was also quarantined at JSC along with the crew. At the end of the quarantine, on August 14, 1969, the capsule was loaded aboard a Super Guppy aircraft at Ellington so it could be shipped to North American Rockwell's facility at Downey, California. North American Rockwell was the prime contractor for the capsule, and Downey was the factory that built them. (NASA.)

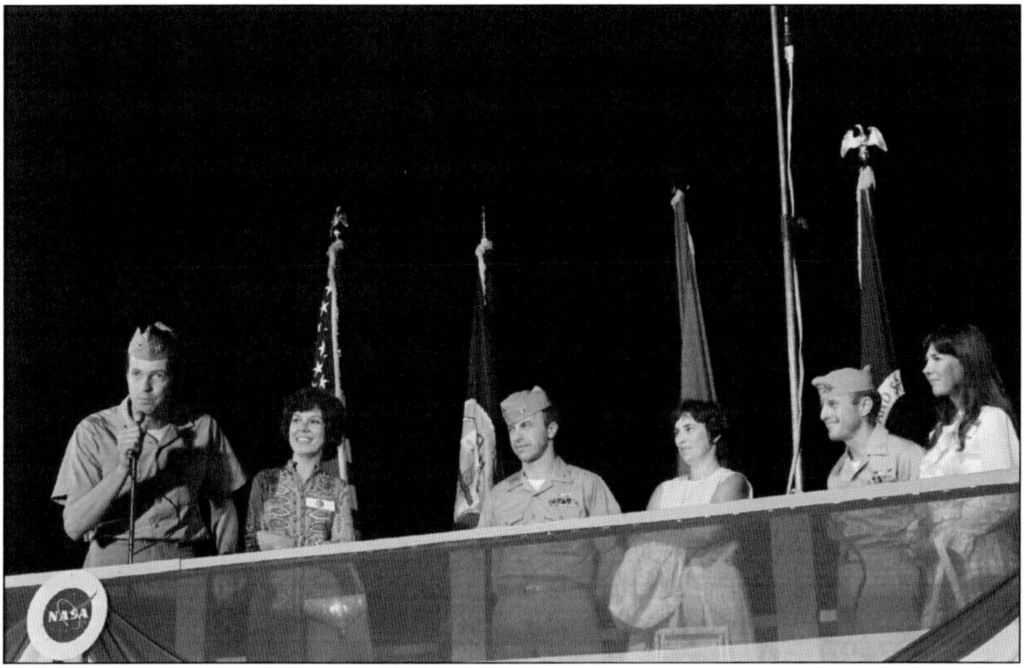

Greeting returning astronauts at Ellington was a tradition that outlived Apollo. Here, the three astronauts who flew the first manned Skylab mission speak to the crowd welcoming them home at the end of their mission in June 1973. The astronauts are accompanied on the podium by their wives. Skylab 2 lasted over 28 days. It was NASA's longest manned mission at that time. (NASA.)

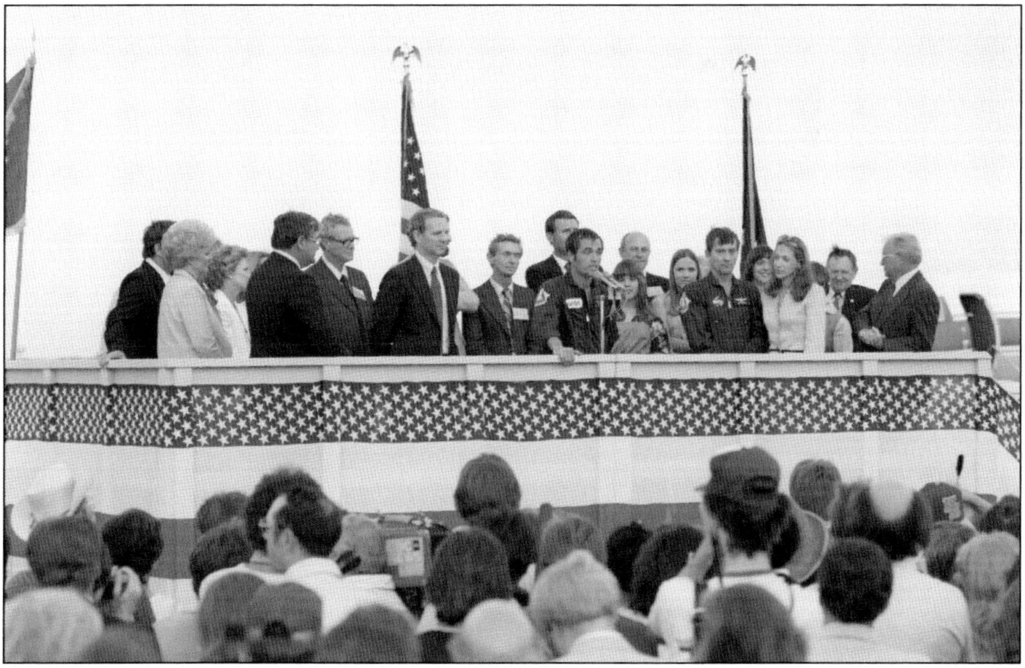

Crowds also greeted every returning shuttle crew at Ellington. This is the very first crew, STS-1's pilot Robert Crippen (left, speaking into the microphone) and commander John Young. They are accompanied by their wives and JSC officials. (NASA.)

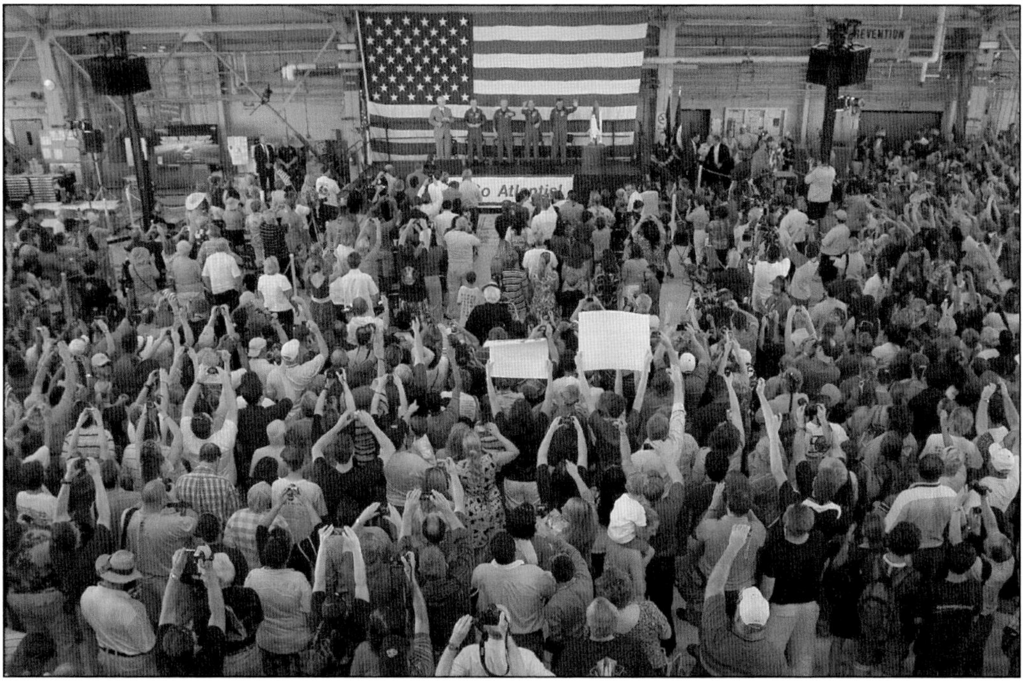

The crew of STS-135, the final shuttle mission, brings down the curtain on the space shuttle program. They are greeted by an enthusiastic crowd in NASA hangar 990 at Ellington Airport. (NASA.)

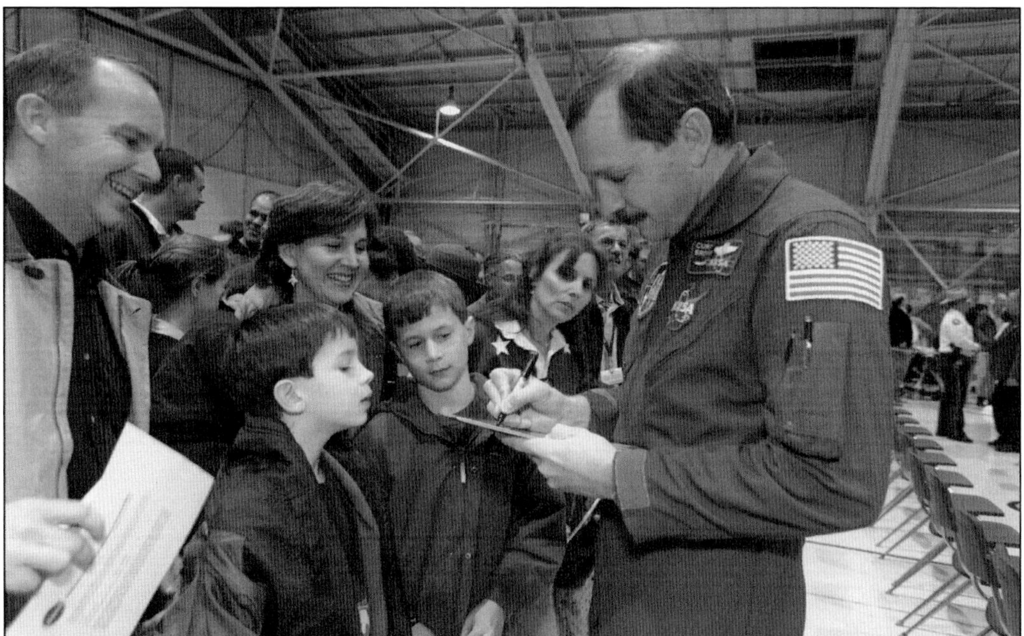

For over 50 years, Ellington has been the place where the astronauts first meet the public at the end of a mission, often signing autographs for admiring spectators. (NASA.)

It is also the place where astronauts first greet their families after returning from space. Here, STS-101 pilot Scott J. Horowitz reunites with his wife, Lisa Marie, and young daughter. (NASA.)

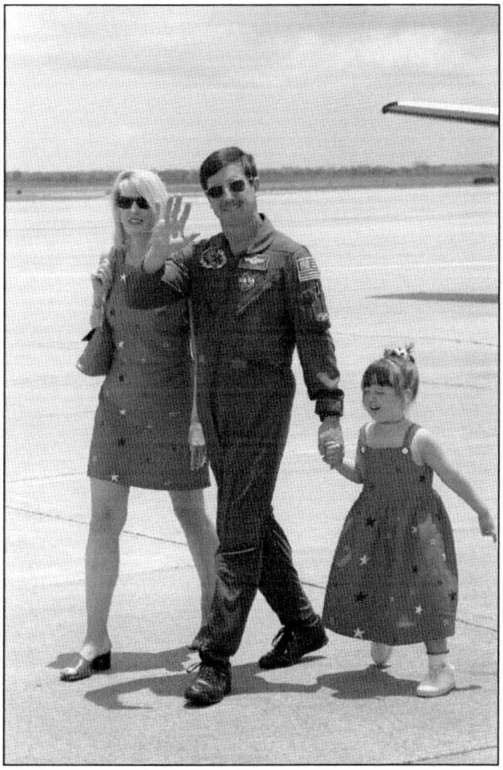

Ellington is also the place where the NASA aircraft assigned to JSC are stationed. These four T-38 Talon trainers, used by the astronauts to maintain flight proficiency and for transport, make a low-level formation pass over Ellington's main runway. On the apron are the KC-135 reduced gravity trainer (left), and a Super Guppy transport. (NASA.)

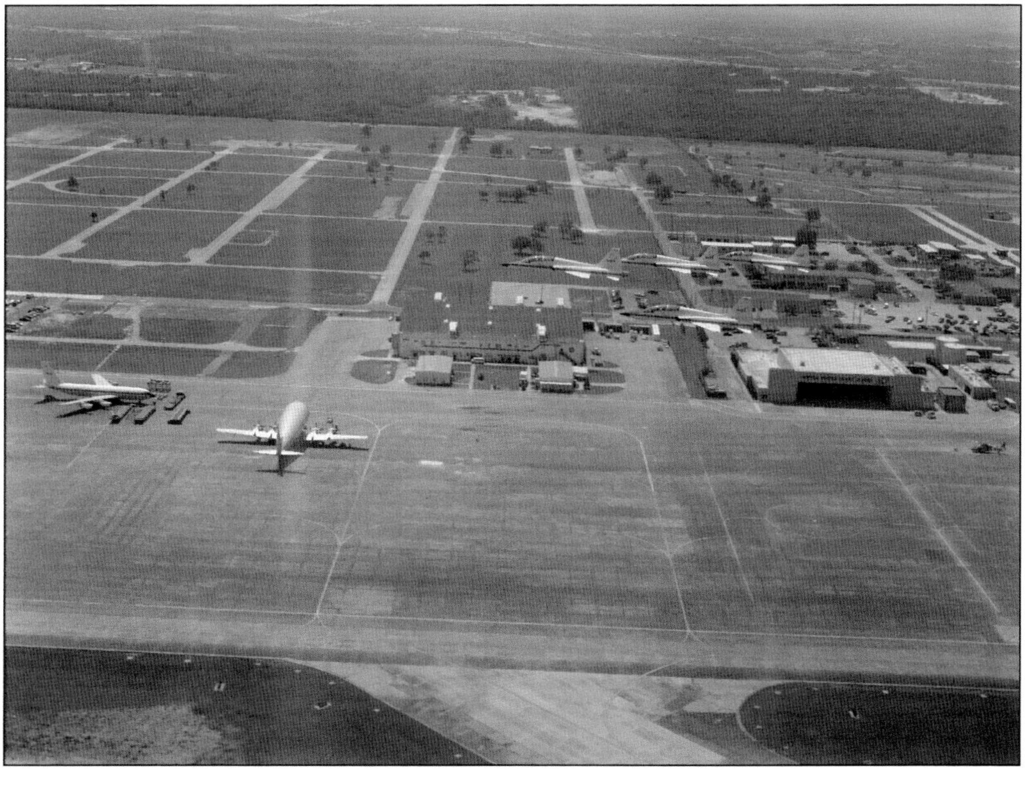

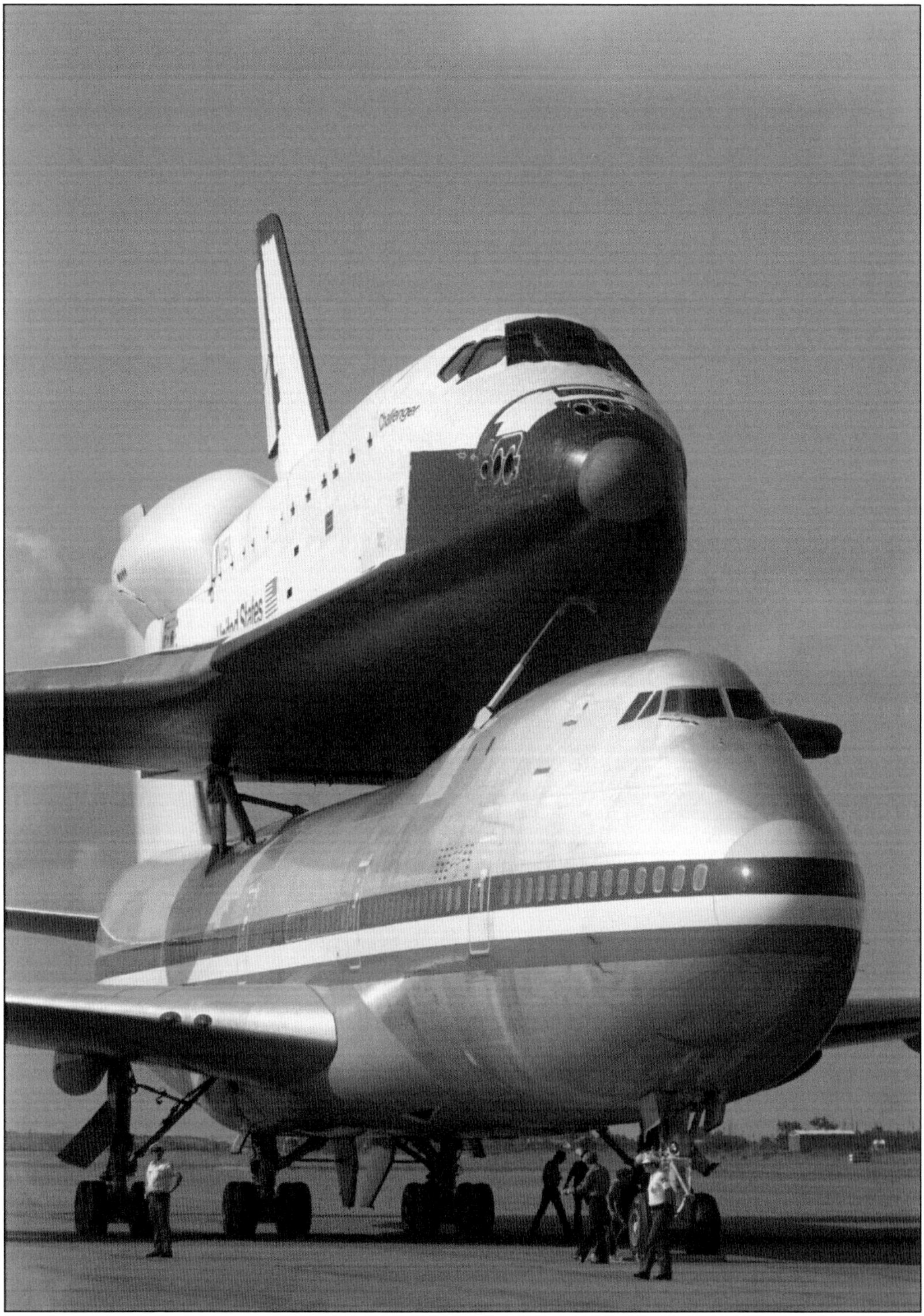

On the infrequent occasions a shuttle orbiter arrived in Houston, it landed at Ellington, carried by a 747 Shuttle Carrier Aircraft. In this photograph, the SCA taxies with orbiter *Challenger* atop it during a 1982 visit to Ellington. (NASA.)

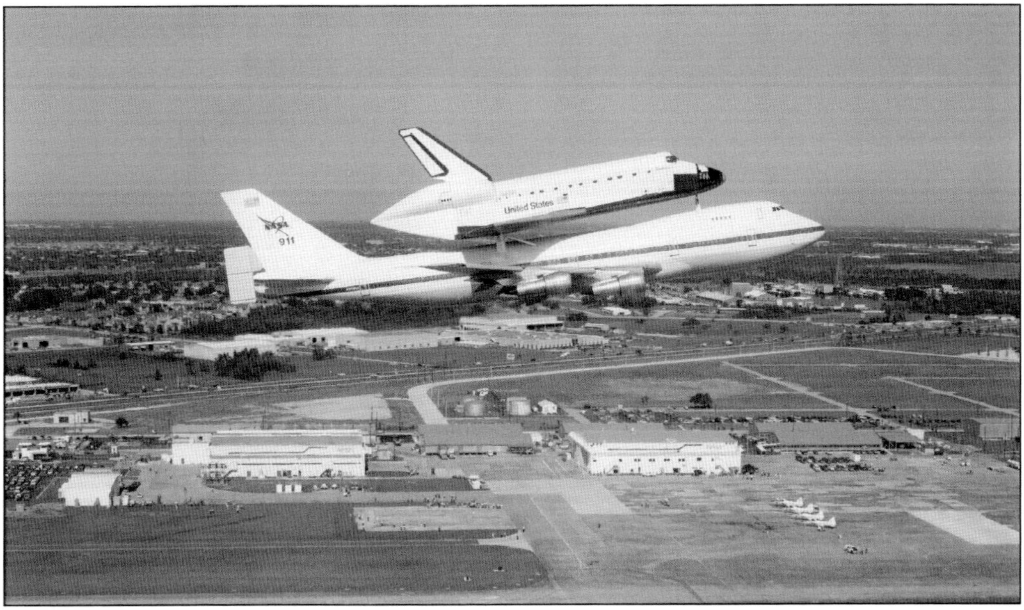

Nine years later, in 1992, the second SCA makes a low pass over Ellington carrying the orbiter *Endeavor*, following a May 1991 visit. The SCA made its final landing at Ellington in 2014, prior to being sent to Space Center Houston to become a museum exhibit. (DVIDS.)

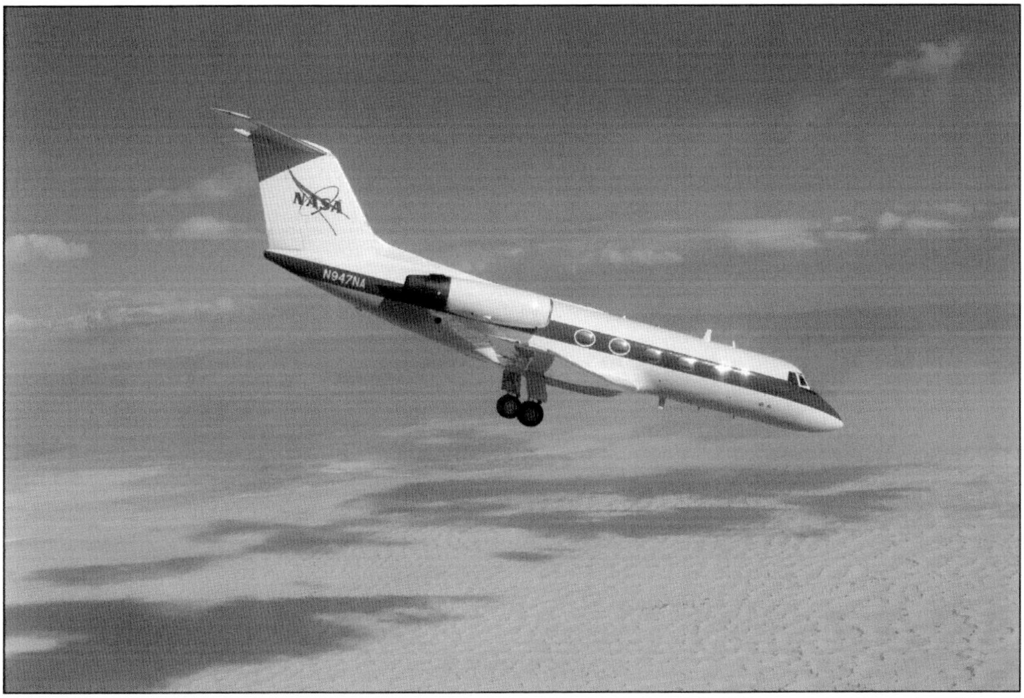

Another NASA aircraft maintained at Ellington during the space shuttle program was the Shuttle Training Aircraft. This was a Grumman C-11A Gulfstream modified to have a shuttle cockpit on the left side and special air brakes that allowed it to mimic the orbiter descent profile. In this configuration it had the sink rate of a brick, just like an orbiter. It was a frequent Ellington visitor. (NASA.)

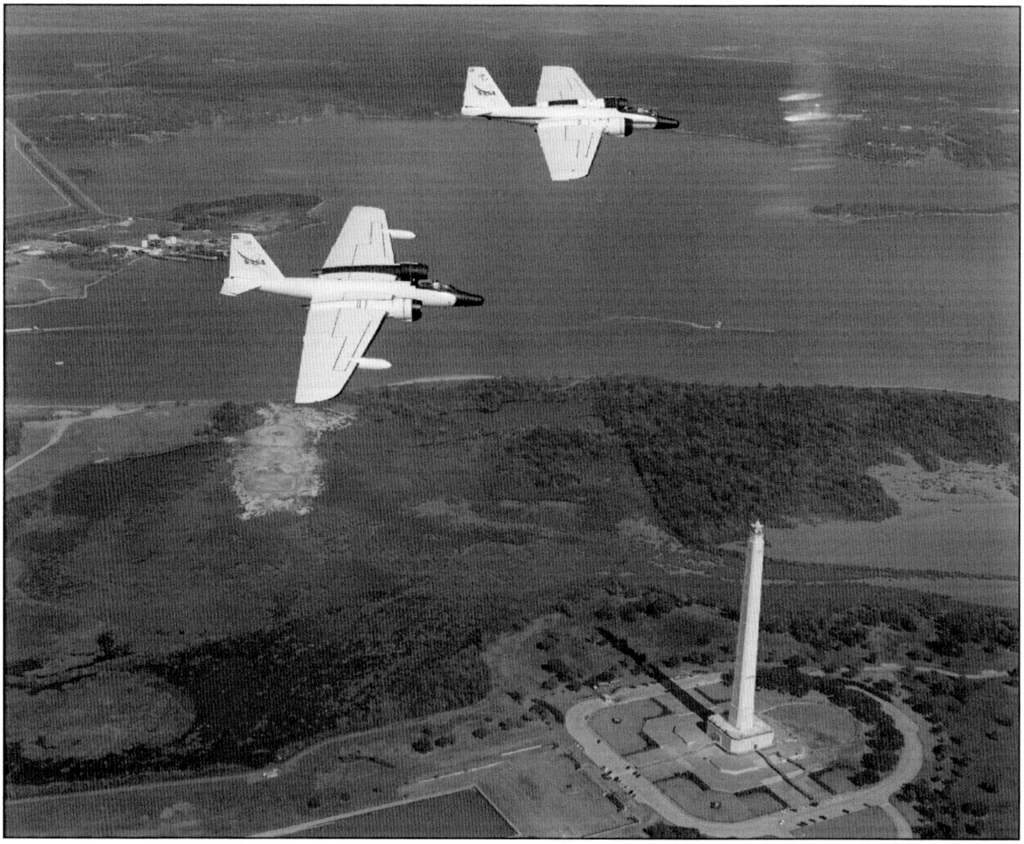

Not all NASA aircraft at Ellington were related to NASA's manned spaceflight role. Ellington is home to three WB-57 aircraft. Two are seen here flying past the San Jacinto monument. Based on a 1950s design, these aircraft are used for high-altitude atmospheric research. Their wings and fuselage give them the appearance of a flying dinosaur. (NASA.)

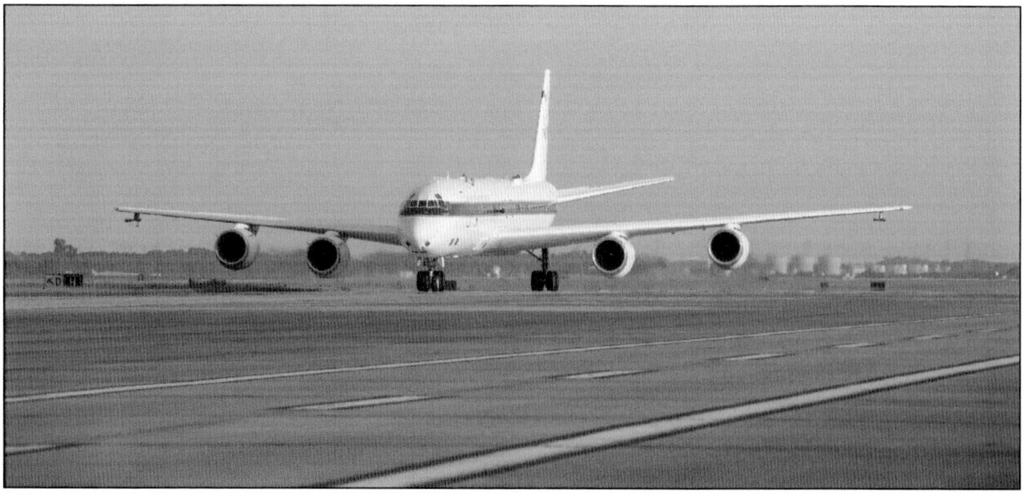

Ellington also hosts NASA aircraft based at other NASA centers when they need to operate out of Houston. This NASA DC-8 is normally based at Palmdale, California. It was at Ellington conducting atmospheric research, measuring pollution over the Gulf of Mexico. (NASA.)

Five

A Third Municipal Airport 1984–2018

As the Air Force lost interest in Ellington during the 1970s, the City of Houston gained interest in acquiring it. For many years, Houston had one major commercial airport: Houston Municipal Airport (today's William P. Hobby Airport) southeast of downtown. It was built before the jet age for twin-engine, tail-dragging airliners like the Boeing 247 and DC-3.

By the late 1950s, it was obvious the existing airport was too small, and in 1965, Houston built a massive new airport on the north side to replace Hobby as Houston's passenger airport. Opened in 1969, it became the George H.W. Bush Intercontinental Airport. Plans to close Hobby to passenger service were dropped when Houston's growth meant Intercontinental would be too crowded without Hobby.

This meant Houston needed a new air cargo hub. The city experienced precipitous growth in the 1960s and 1970s. In 1981, the Reagan administration considered Ellington as a site to house Cuban refugees escaping to the United States from the Mariel Boat Lift. The World War II barracks were too deteriorated to use, however. The plan caused Houston's City Council to consider acquiring the airbase.

On July 1, 1984, the City of Houston took possession of Ellington. NASA and the US armed forces retained possession of their ground facilities at the renamed Ellington Airport, but the Houston Airport System took responsibility for maintaining the runways, air traffic control facilities, streets, taxiways, and remaining civilian buildings on the airport. The Air Force and Air National Guard became the Ellington Field Joint Reserve Base on the northwest portion of the airport. It houses Air Force, Army, and Navy reserve units. Additionally, the US Coast Guard operates out of Ellington.

Over the next 30 years, the HAS demolished and replaced the dilapidated buildings remaining from World War II, refurbished Cold War–era buildings, and erected new facilities to serve the commercial and general aviation market at Ellington. Commercial passenger operations were mostly commuter flights to Intercontinental, which ceased in 2004. Ellington has proved invaluable for general aviation and a focus for disaster relief operations in the Houston area.

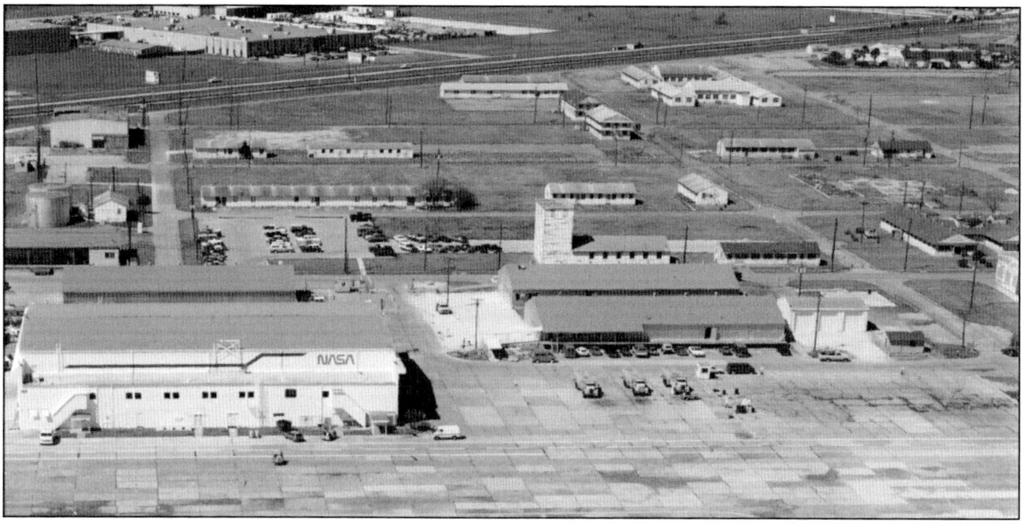

When the Houston Airport System took over Ellington, many of the World War II–era wooden offices and two-story barracks buildings still existed. These would be torn down over the next decade due to their poor condition. (NASA.)

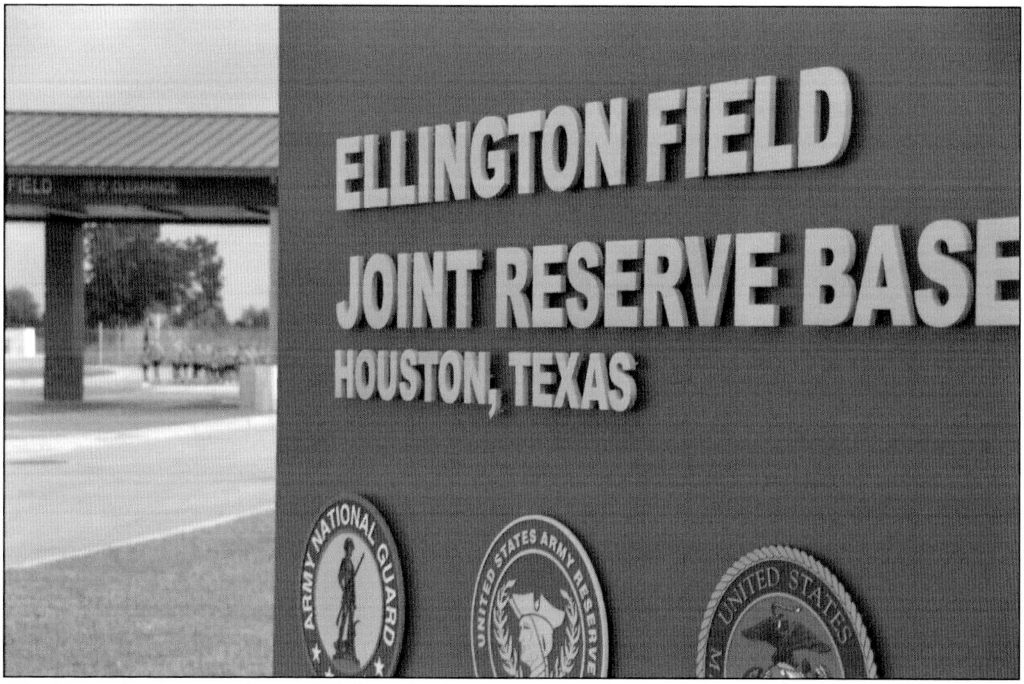

The Texas Air National Guard did not leave Ellington, however. They remained, with the military presence at Ellington expanded. The base became the Ellington Field Joint Reserve Base, with the Texas Air National Guard, Texas Army National Guard, Army Reserve, Navy Reserve, and Marine Corps Reserve operating out of Ellington. (DVIDS.)

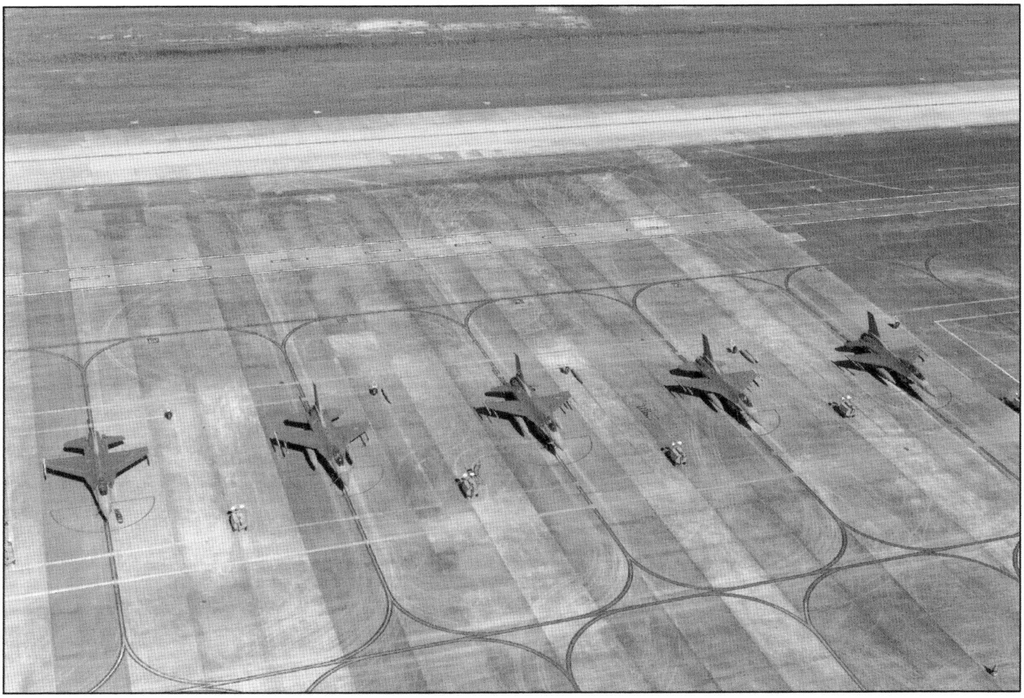

The Texas Air National Guard was reequipped with F-16s in 1989, replacing the F-4s previously operating out of Ellington. These aircraft are on alert on the apron at Ellington. (LOC.)

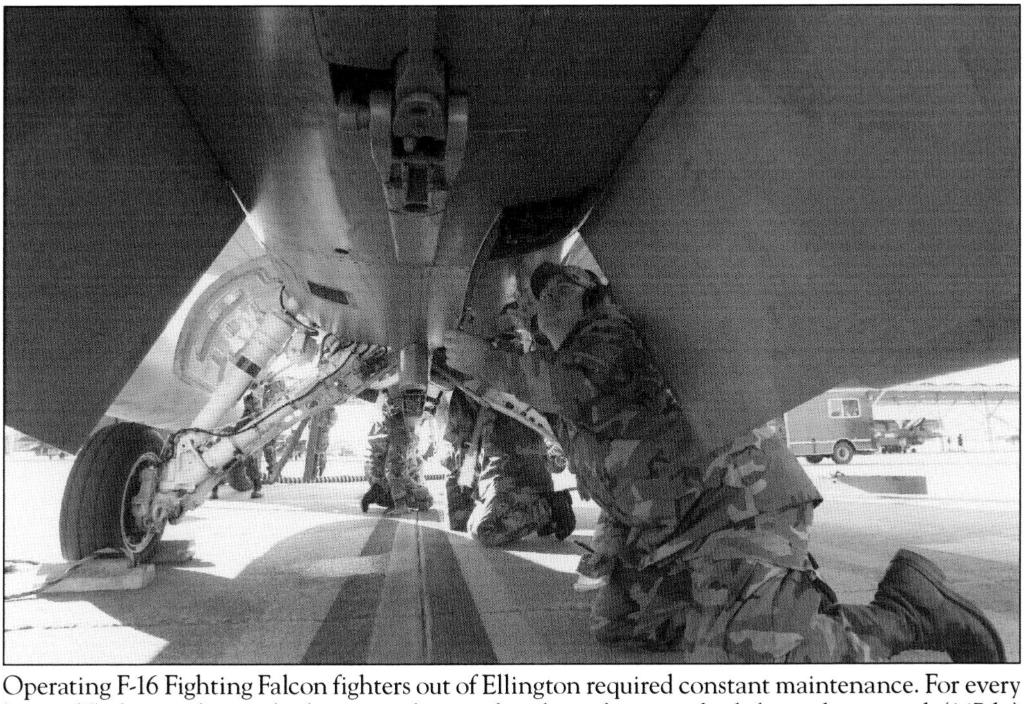

Operating F-16 Fighting Falcon fighters out of Ellington required constant maintenance. For every hour of flight, mechanics had to put in hours of work on the aircraft while on the ground. (147th.)

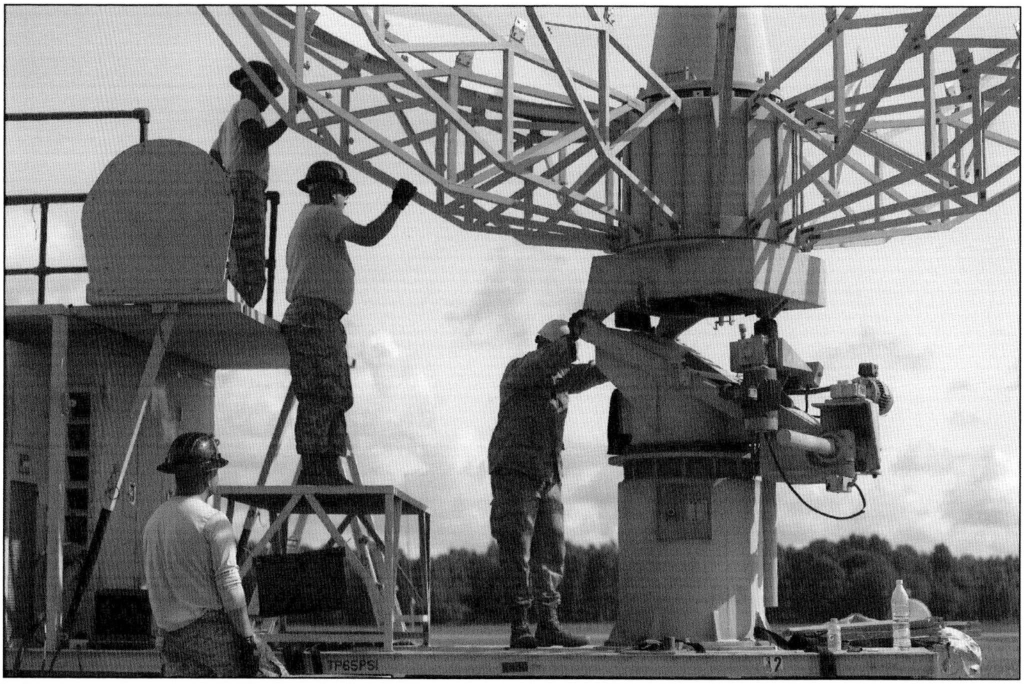

The Texas Air National Guard also needed to ensure the radars at Ellington worked. Here, TANG personnel service a radar unit at Ellington Field in the 1990s. (147th.)

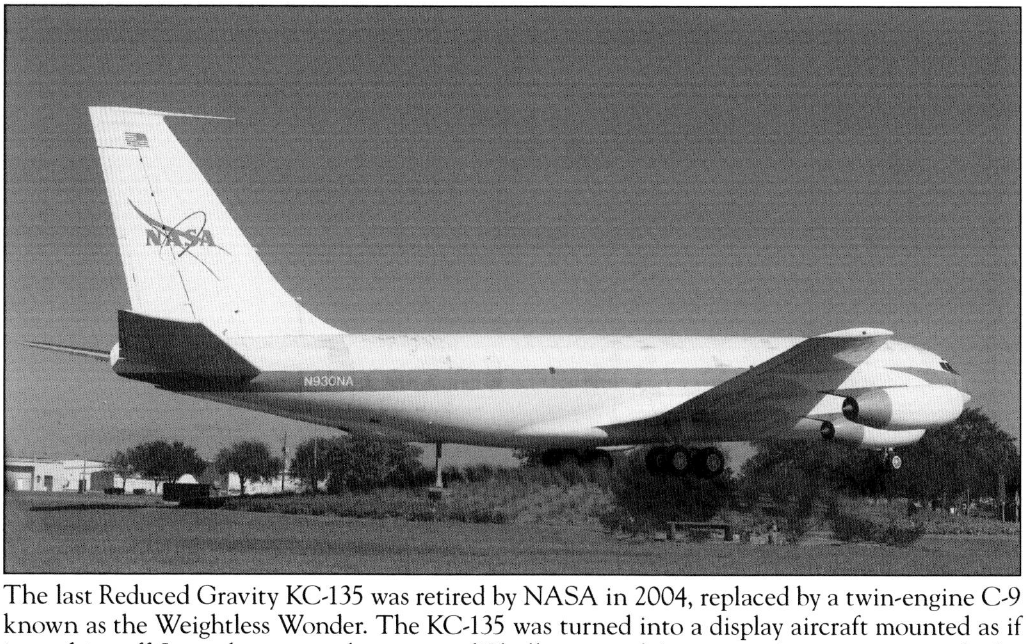

The last Reduced Gravity KC-135 was retired by NASA in 2004, replaced by a twin-engine C-9 known as the Weightless Wonder. The KC-135 was turned into a display aircraft mounted as if just taking off. It can be seen at the corner of Challenger and Aerospace Avenues, just inside the main entrance to Ellington Airport. (LOC.)

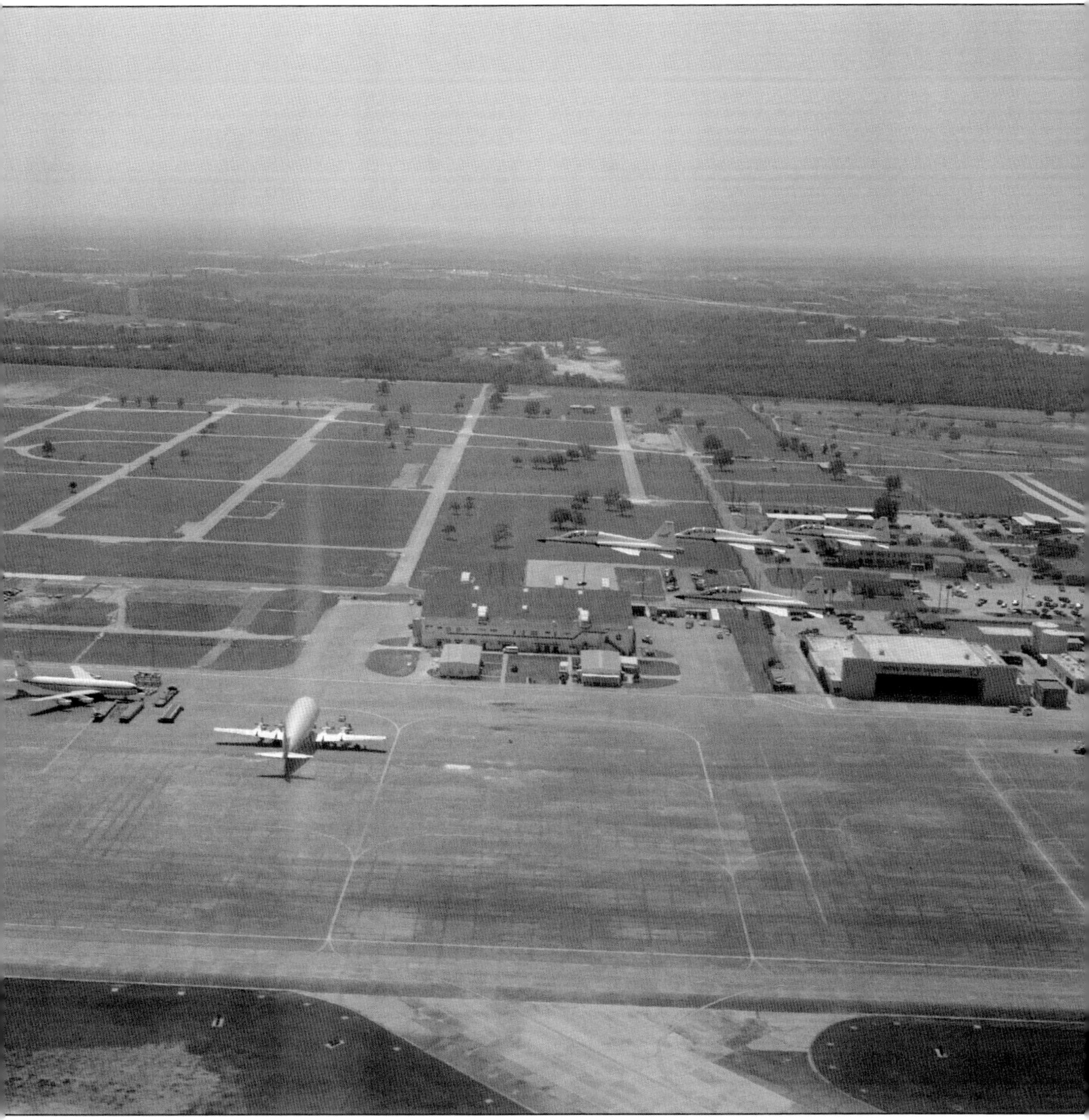

NASA also maintained its presence at Ellington after the City of Houston took over the airfield. This photograph shows four NASA T-38s making a formation pass over the airfield. On the ground are NASA's Reduced Gravity KC-135 (left) and one of NASA's Super Guppy cargo aircraft (center). (DVIDS.)

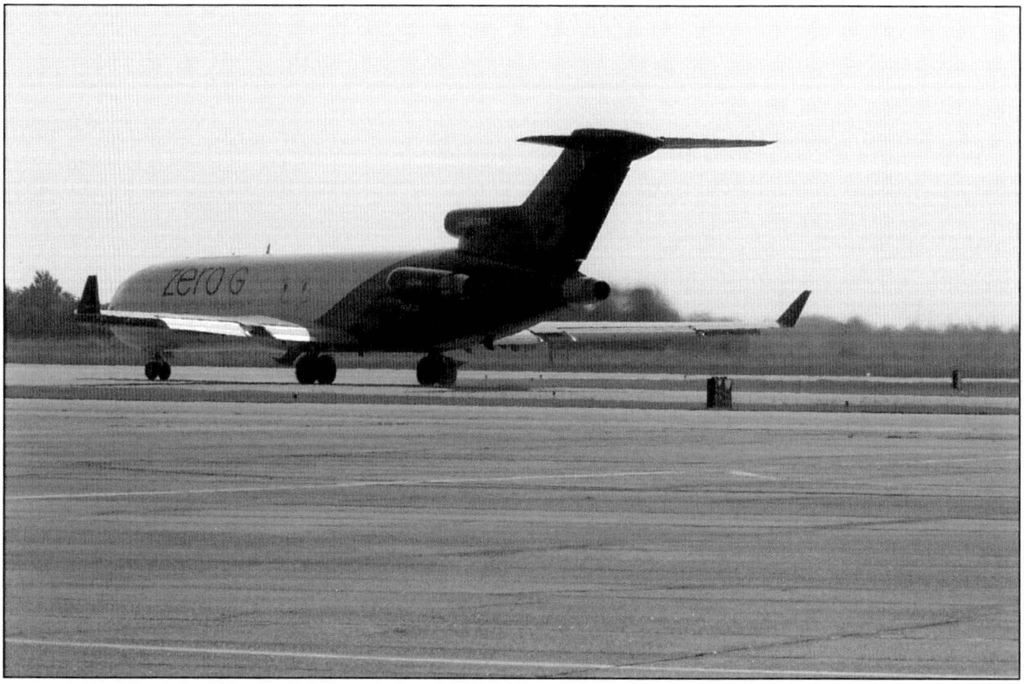

Commercial opportunities to experience zero gravity now exist, and a seat aboard one of Zero-G Corporation's aircraft can be purchased by anyone with the price of a ticket. Founded in 1994, Zero-G Corporation schedules flights out of Ellington. (DVIDS.)

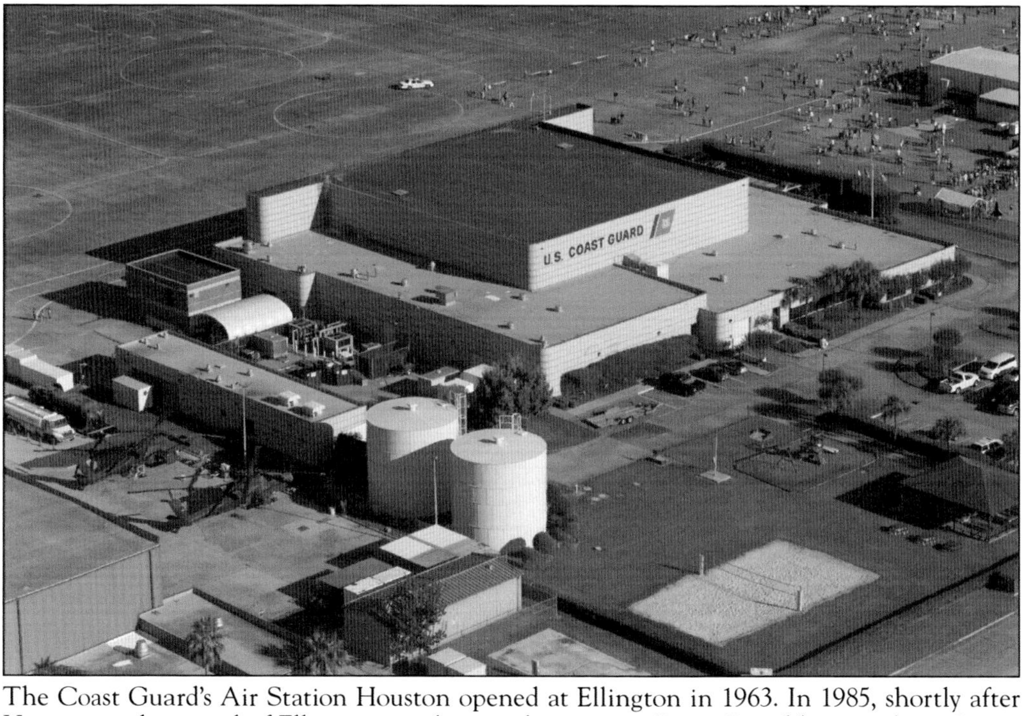

The Coast Guard's Air Station Houston opened at Ellington in 1963. In 1985, shortly after Houston took control of Ellington, work started on a new Coast Guard hangar there. It was completed in late 1986. (DVIDS.)

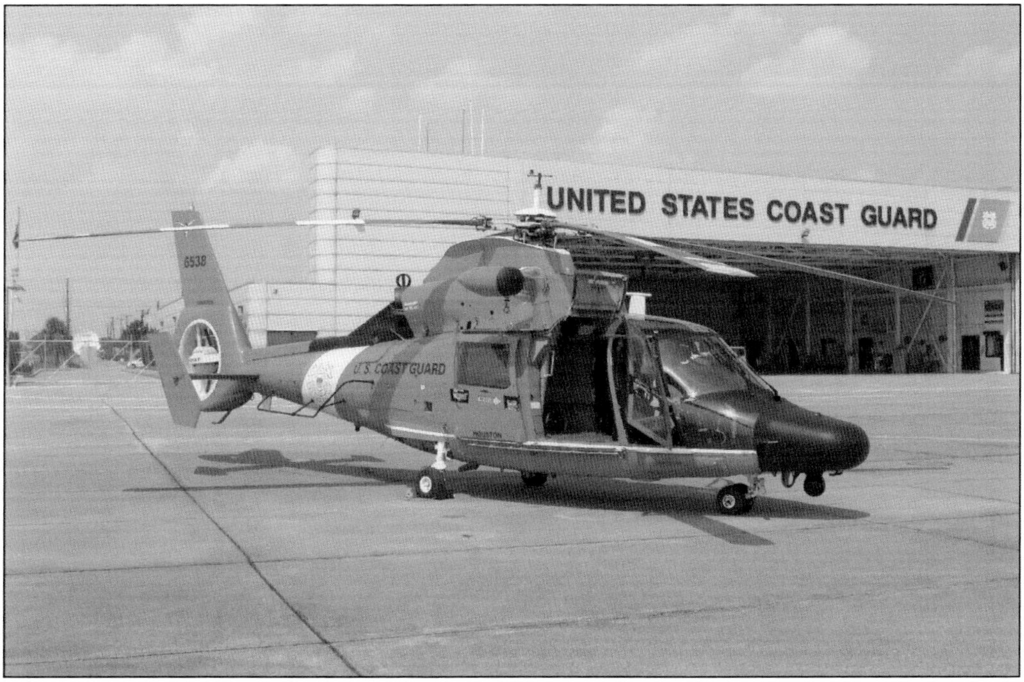

The original aircraft stationed at Coast Guard Station Houston were Sikorsky HH-52 Seaguard helicopters. By the time HAS took charge of Ellington, the Seaguards had been replaced by the newer, faster, and nimbler American Eurocopter MH-65 Dauphin. (USCG.)

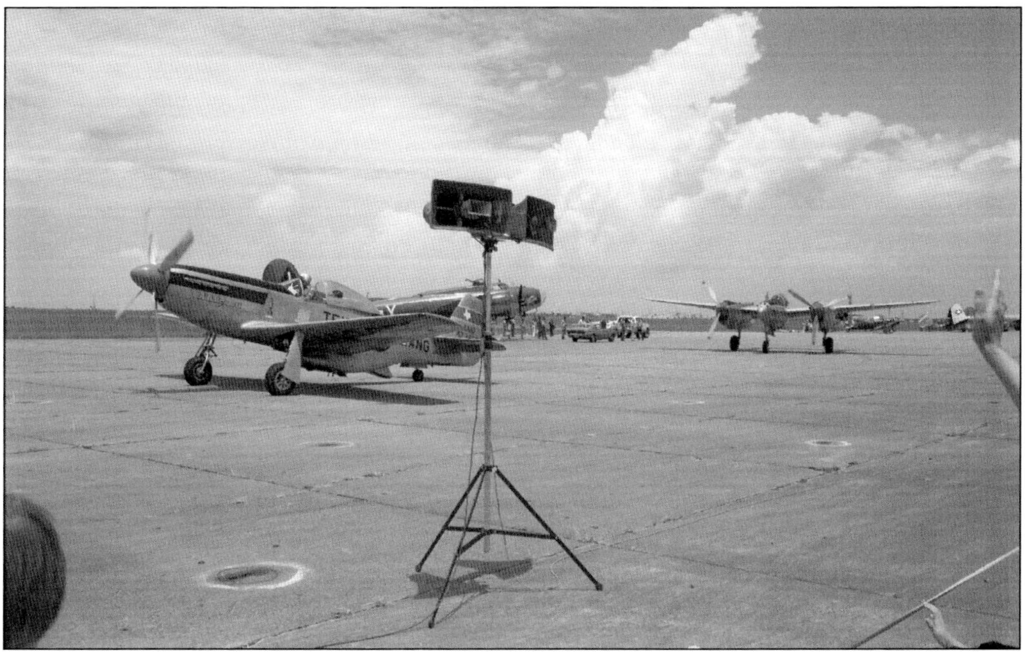

Because Ellington has relatively little commercial traffic but is capable of handling high-performance aircraft, it is a natural location for hosting air shows. One of the earliest was the 1969 Gathering of Eagles. Here, World War II–era warbirds parade on Ellington's apron. In the foreground is an F-51 in Air National Guard colors. Behind it is a B-17. To the right is a P-38 Lightning. (Shawn Harrison.)

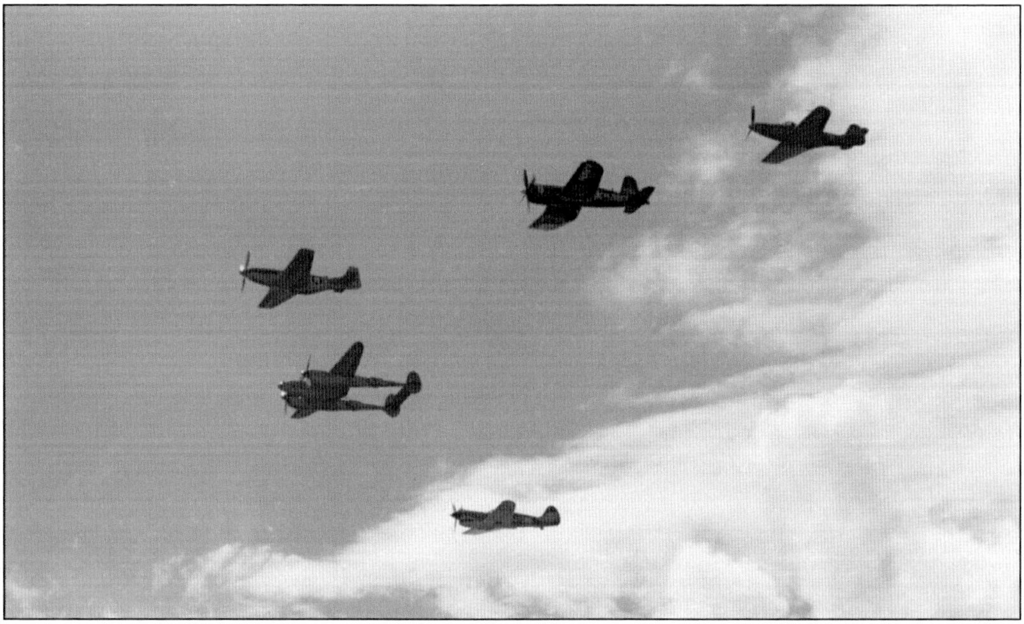

That show featured aircraft flybys. This photograph shows World War II US fighters in a "V" formation. Leading is an F-51 Mustang in Air National Guard markings. (In World War II, it was a P-51.) From left to right are a P-38 Lightning, a P-40 Warhawk in American Volunteer Group (Flying Tigers) colors, an F-4U Corsair in US Navy markings, and a P-63 King Cobra painted in Soviet colors. (Shawn Harrison.)

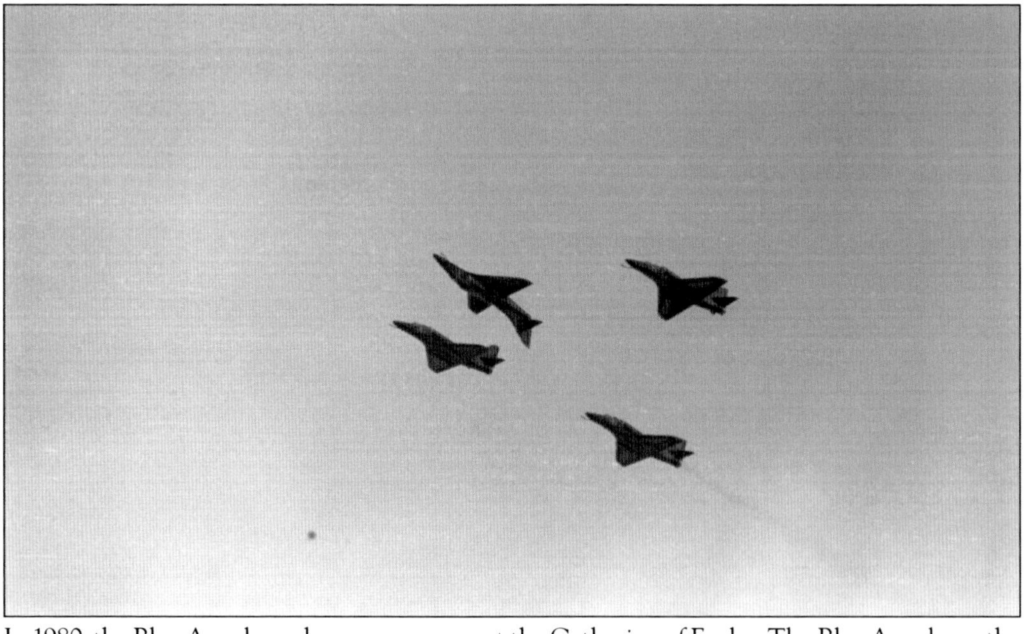

In 1980, the Blue Angels made an appearance at the Gathering of Eagles. The Blue Angels are the Navy's flight demonstration team. At the time, they were flying A-4 Skyhawks, a light bomber. Ever since then, Ellington air shows have featured appearances by precision demonstration flying teams, generally the Blue Angels or Air Force Thunderbirds, but sometimes appearances have been made by Canada's Snowbirds or other nations' teams. (Shawn Harrison.)

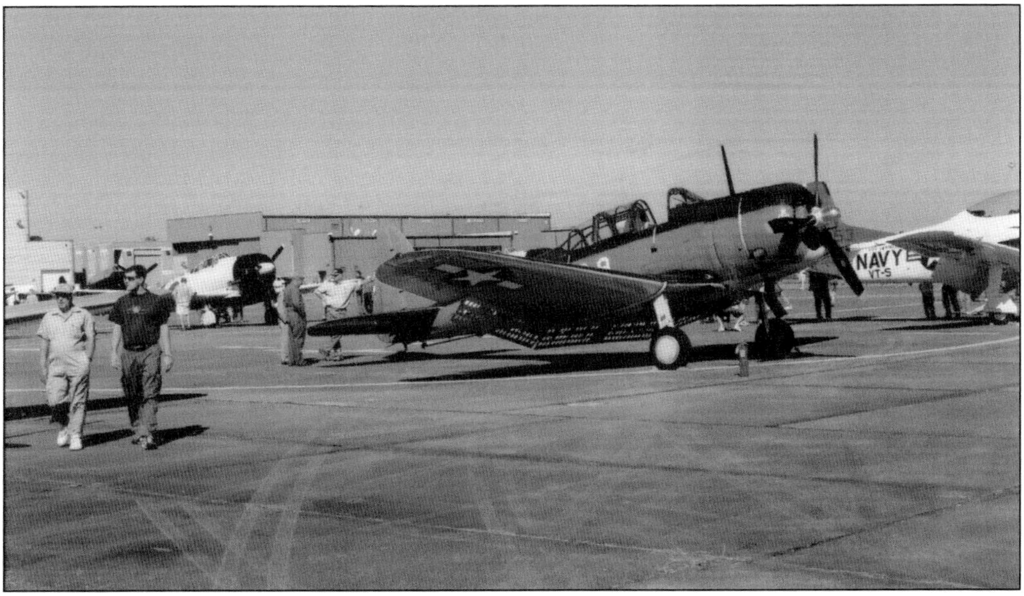

Since 1985, Ellington has hosted the Annual Wings Over Houston air show. It always includes a ground display of interesting aircraft. The World War II warbirds that fly in, such as this Douglas Dauntless, always attract attention, but the show also includes displays of modern military aircraft flown in by the owning countries, and increasingly, 1950s, 1960s, and 1970s warplanes owned by private enthusiasts. (AP.)

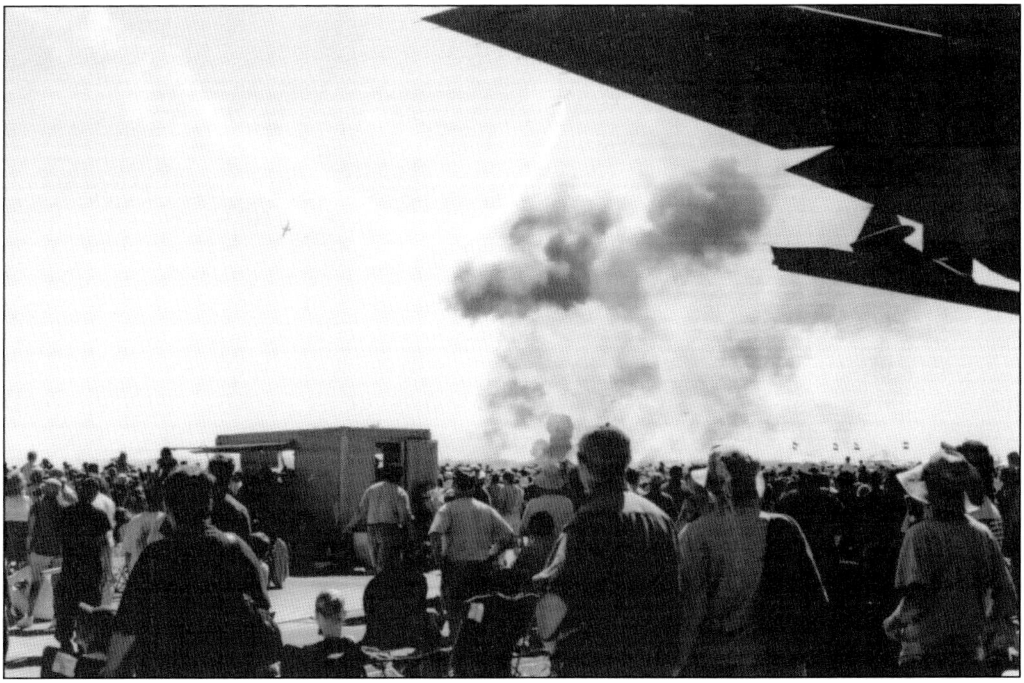

Wings Over Houston always includes air displays, including aerobatics, shows by the Blue Angels or Thunderbirds, and precision parachuting. Every show also includes, as shown here, a simulated recreation of the bombing of Pearl Harbor by replica Imperial Japanese Navy aircraft. A B-17 and P-40 frequently appear during this part of the air show. (AP.)

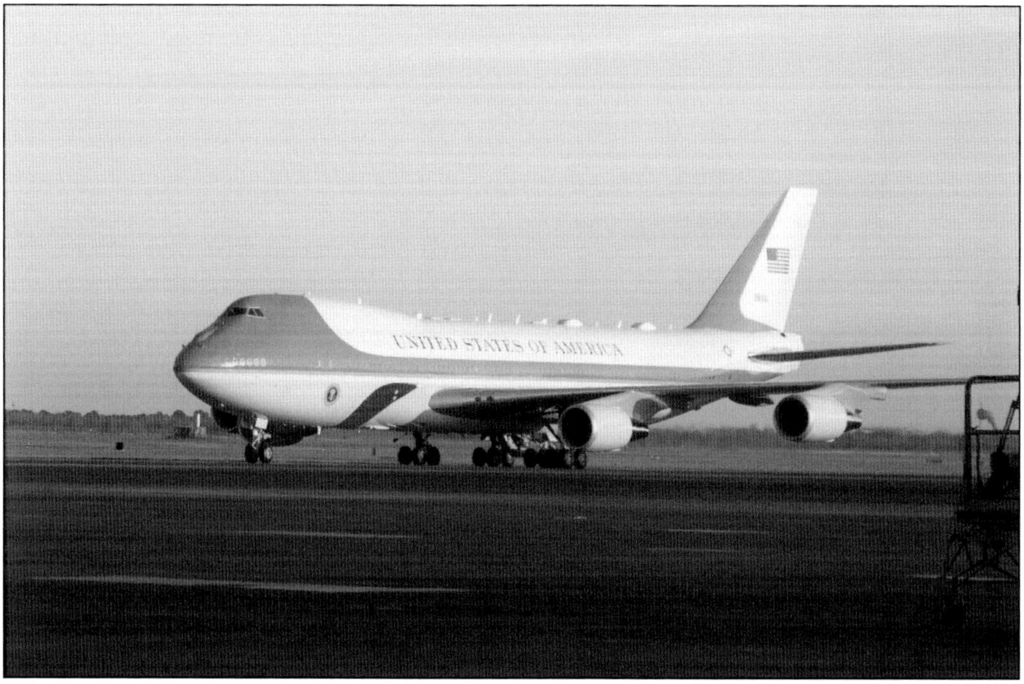

Ellington is also the airport used by visiting dignitaries. If the president flies into Houston, Air Force One, shown taxiing to Ellington's apron, lands at Ellington. (DVIDS.)

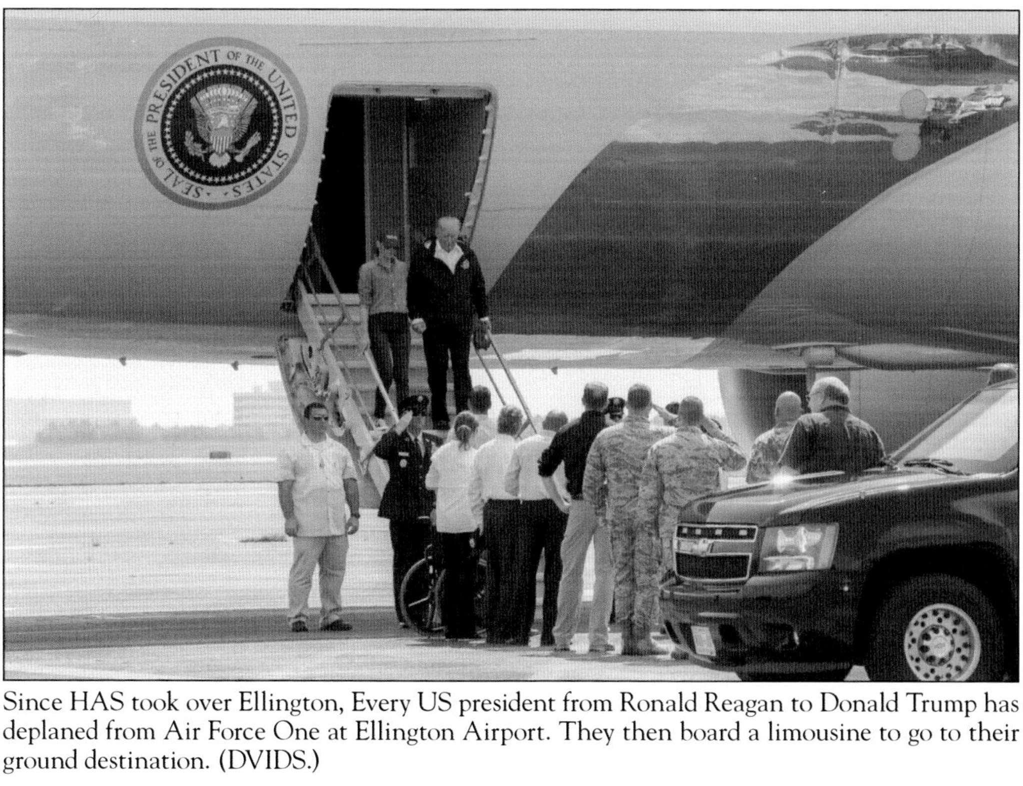

Since HAS took over Ellington, Every US president from Ronald Reagan to Donald Trump has deplaned from Air Force One at Ellington Airport. They then board a limousine to go to their ground destination. (DVIDS.)

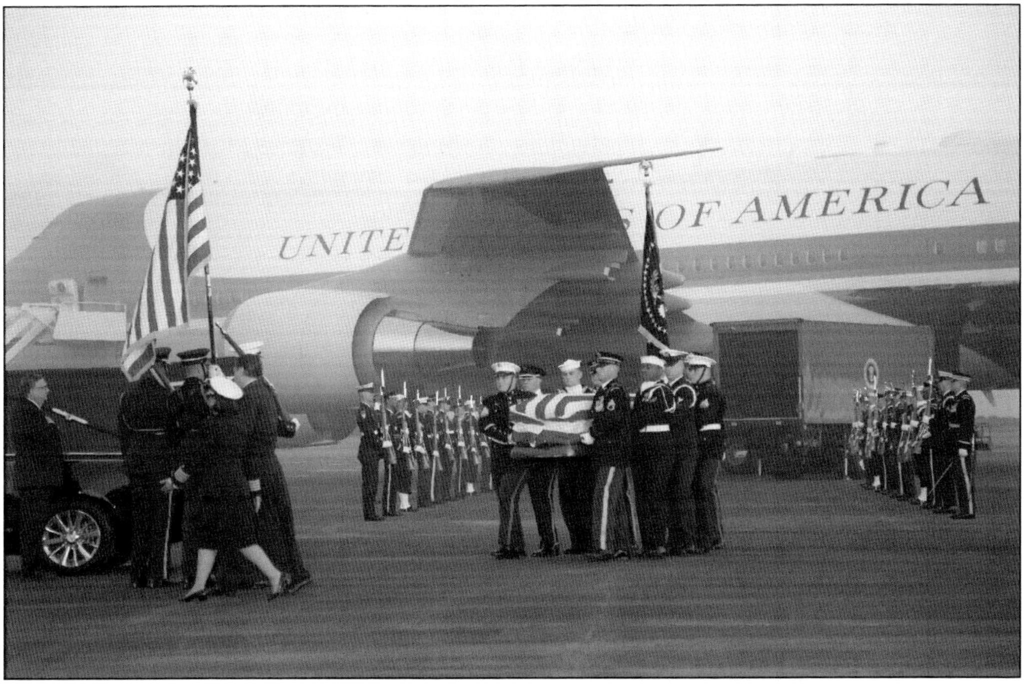

After the death of Pres. George H.W. Bush, Ellington was the final stop for Air Force One with the former president's body. After lying in state at the Capitol Rotunda, the body returned from Washington, DC, for a funeral in Houston. An honor guard turned out as the body was conveyed with due ceremony from Air Force One to the waiting hearse. (DVIDS.)

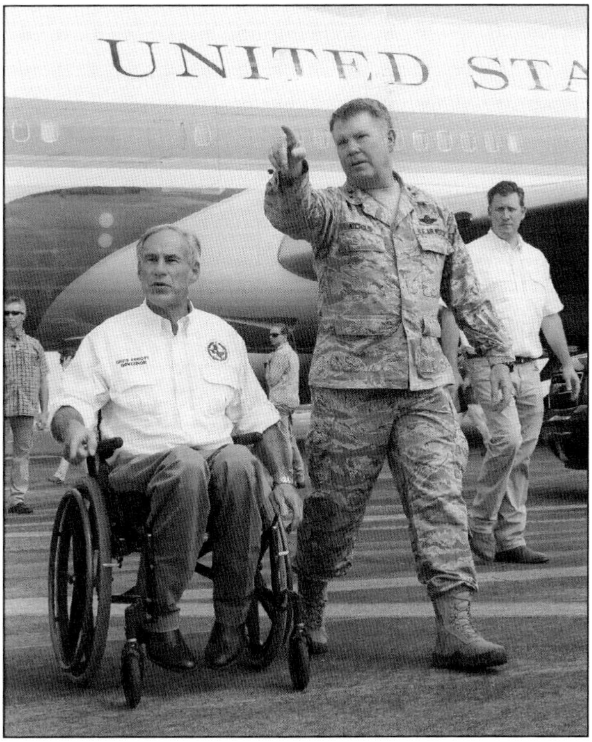

Other officials also use Ellington as their airport on official business. This is Texas governor Greg Abbott at Ellington when Abbott came to Houston to visit areas damaged by Hurricane Harvey. (DVIDS.)

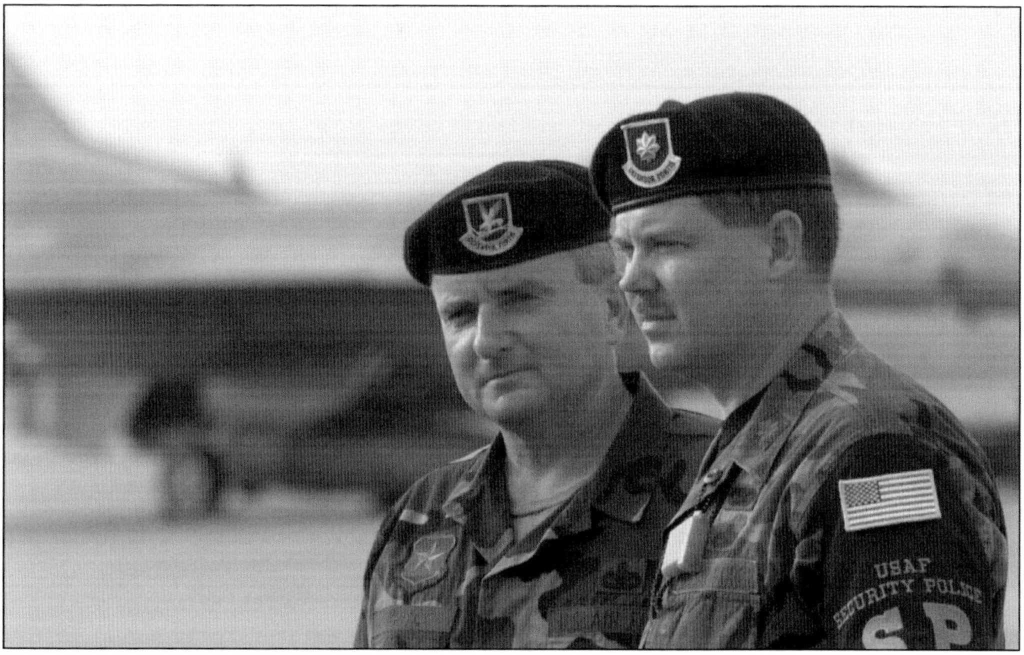

Ellington's military significance became more important in the wake of the September 11, 2001, attacks. Suddenly, homeland security was a priority. The 147th Wing, with the 111th Fighter Squadron, had a new role to play in the defense of the United States; security at Ellington abruptly increased. (147th.)

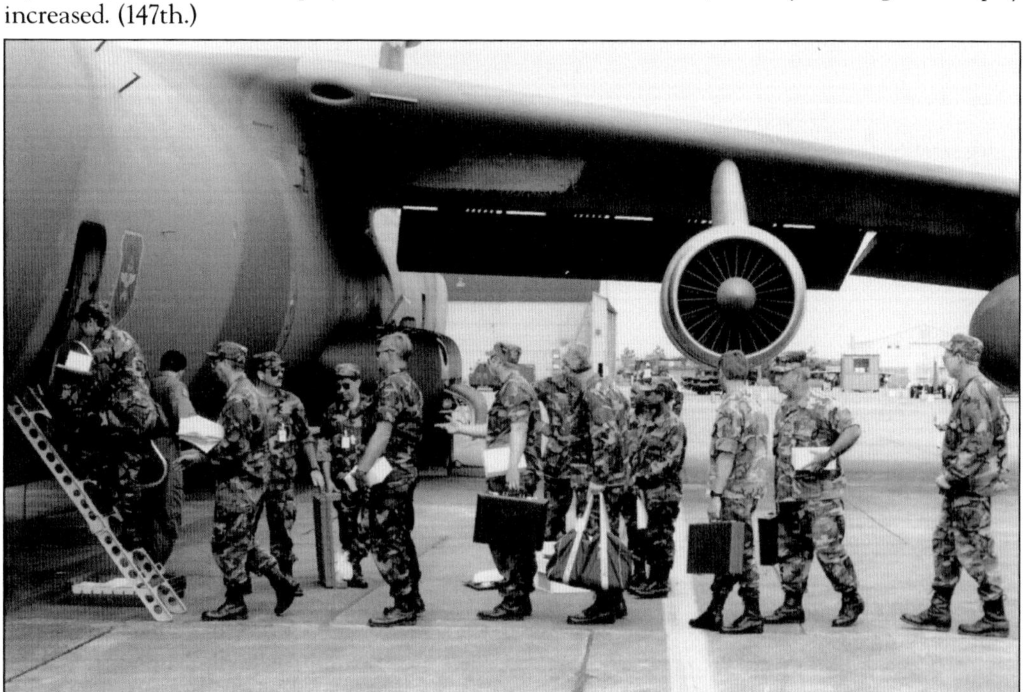

While commercial passenger traffic dropped to almost nothing before 2001, Ellington became an embarkation point during the War on Terror. Thousands of Houston-area soldiers and airmen boarded transports at Ellington to head overseas to support activities in Iraq and Afghanistan. (147th.)

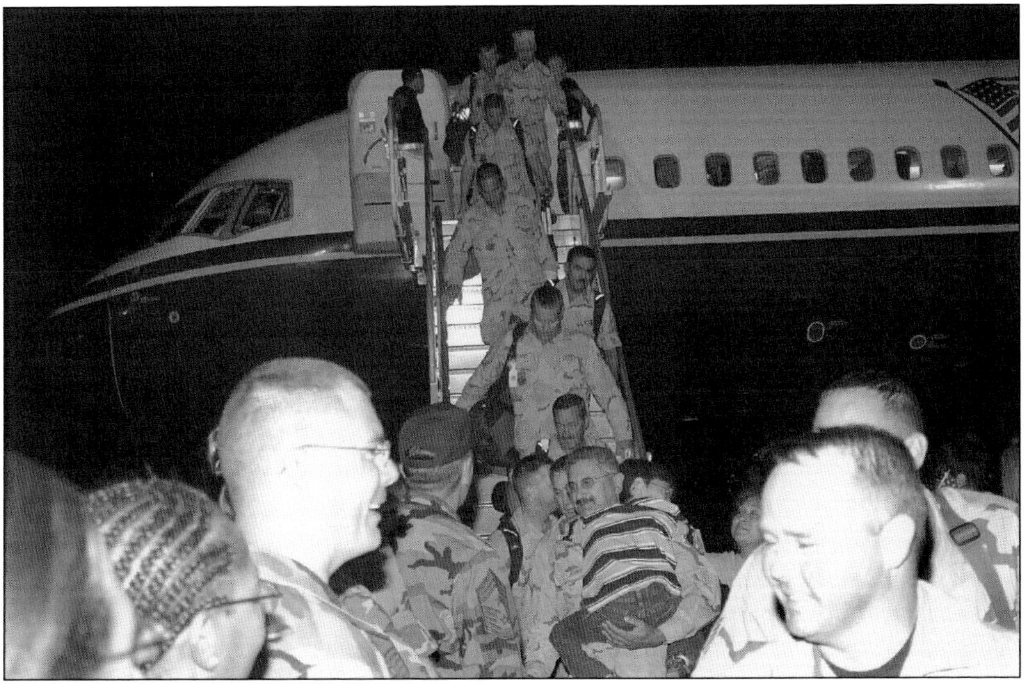

Ellington was also the place where troops returned to the United States after their tours of duty overseas. These 147th Wing airmen debark from an aircraft that left their deployment base at Balad Air Base, Iraq, several hours earlier. (DVIDS.)

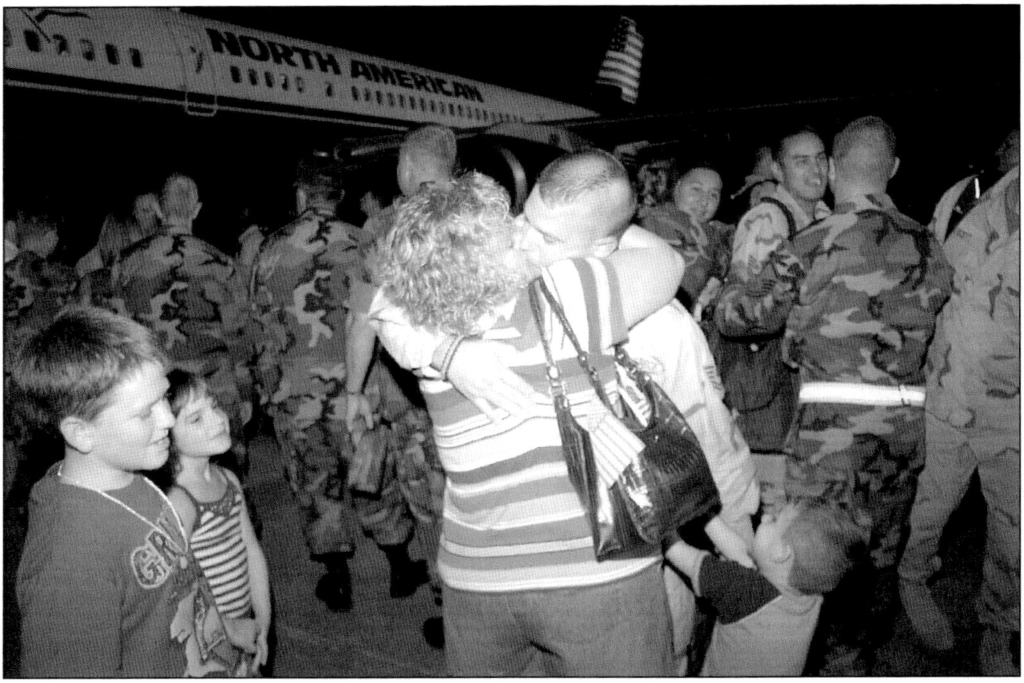

The best part of any return is the chance to reunite with family. Right off the airplane, a husband kisses his wife. Their children put up with the gooey stuff as they await their turn to greet their father. (DVIDS.)

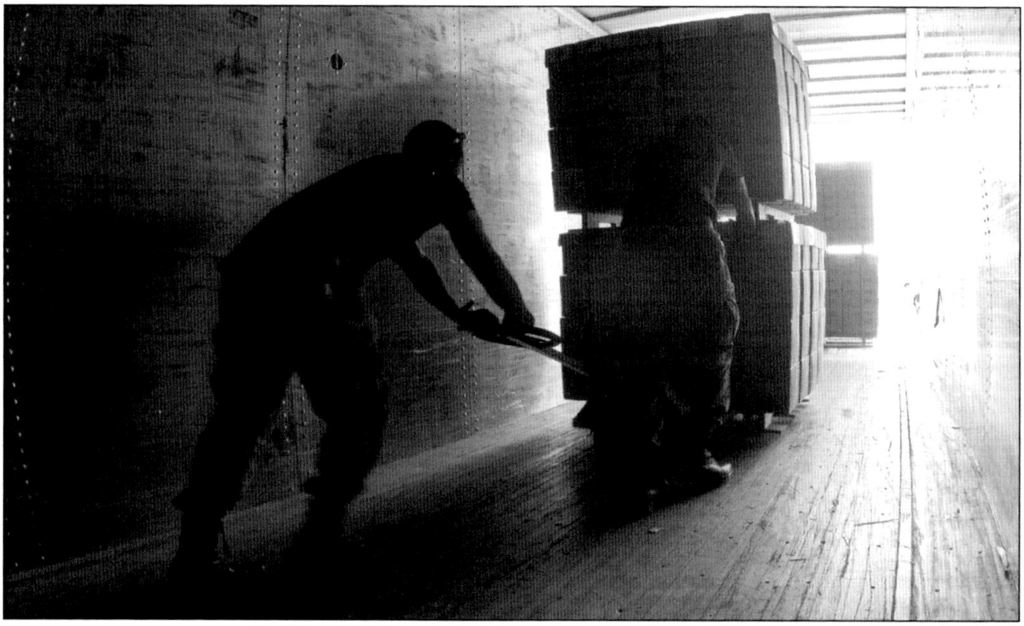

Ellington also provided facilities for hurricane and disaster relief, a traditional role for the National Guard. In this case, Ellington was being used to ship supplies to areas hit by Hurricane Rita, which came ashore near Beaumont, Texas. The two airmen are pushing a pallet filled with bottled water and meals ready to eat (MREs) onto a CH-47 Chinook helicopter, so they can be taken to areas inaccessible by road. (DVIDS.)

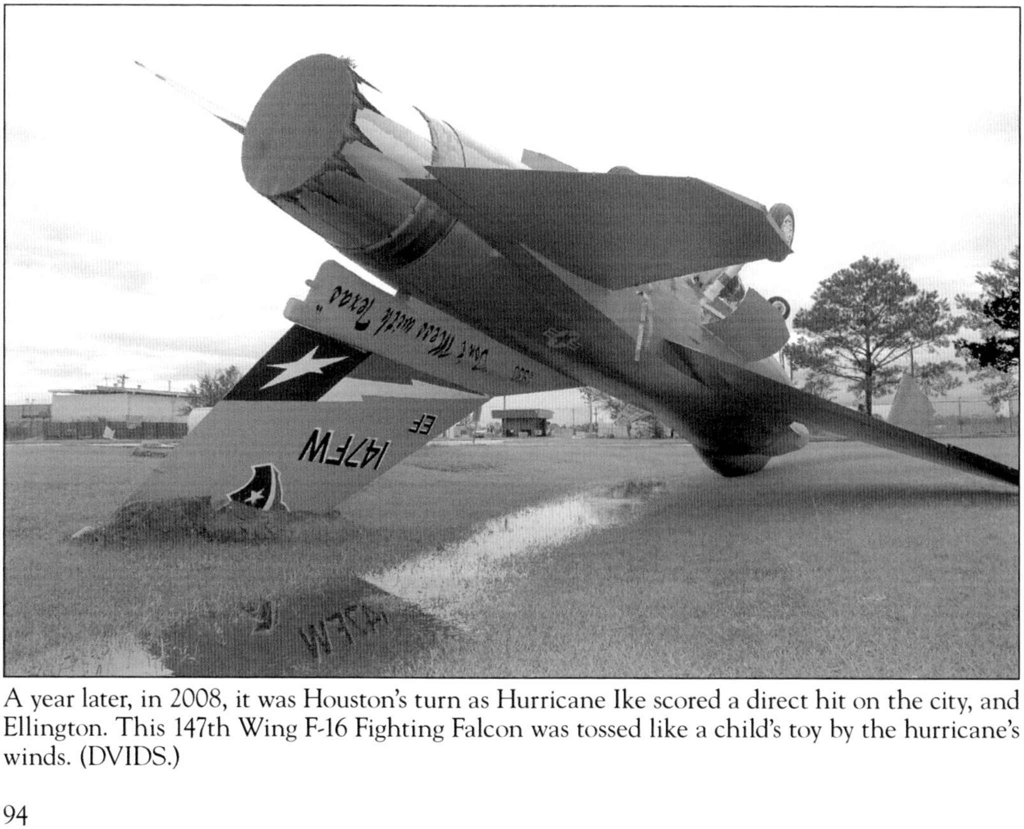

A year later, in 2008, it was Houston's turn as Hurricane Ike scored a direct hit on the city, and Ellington. This 147th Wing F-16 Fighting Falcon was tossed like a child's toy by the hurricane's winds. (DVIDS.)

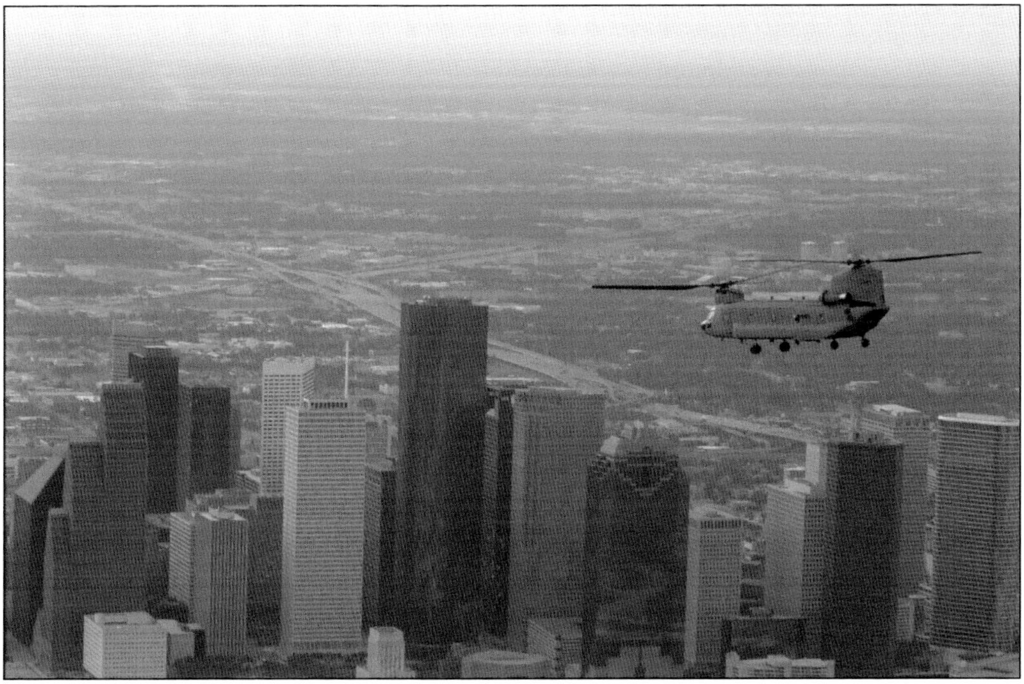

This time, the Chinook helicopters were taking relief supplies to Ellington. This dramatic shot of an Army Chinook over downtown Houston shows it headed outbound for another load of supplies, having unloaded the previous load at Ellington. (DVIDS.)

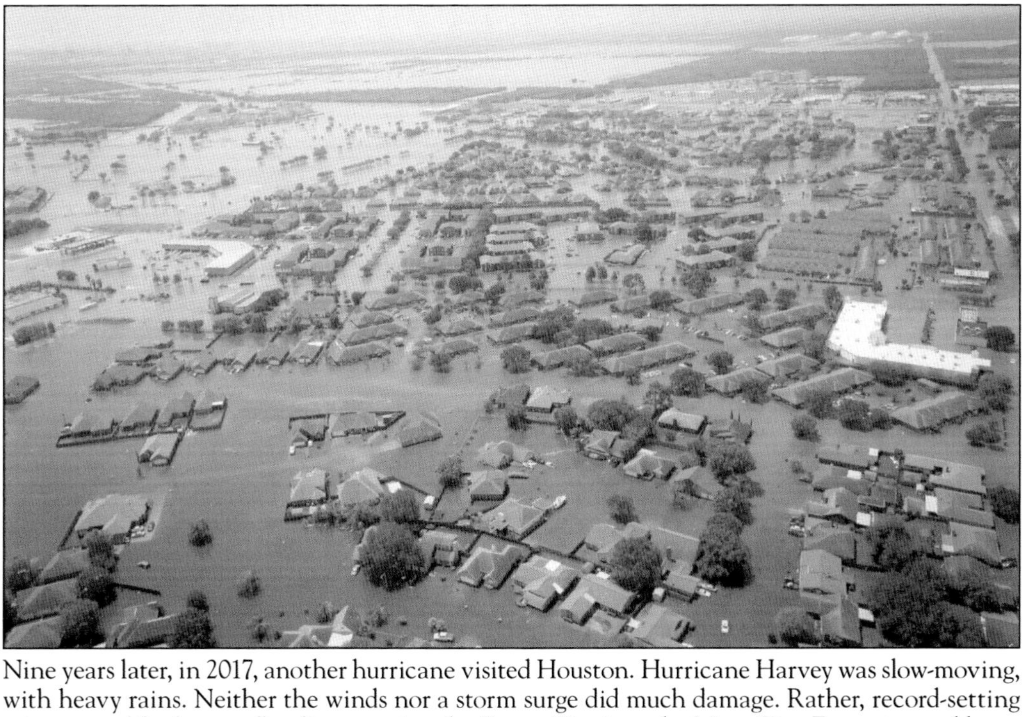

Nine years later, in 2017, another hurricane visited Houston. Hurricane Harvey was slow-moving, with heavy rains. Neither the winds nor a storm surge did much damage. Rather, record-setting rains caused freshwater flooding, turning the Bayou City into the Moat City. Extensive and long-lasting flooding resulted. (DVIDS.)

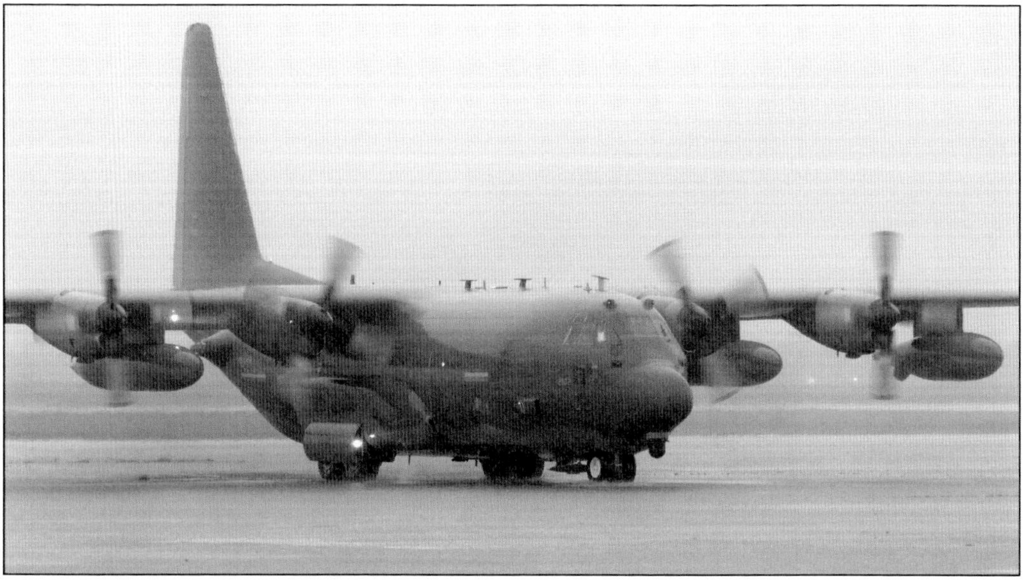

Help came from all over. This photograph shows a C-130 Hercules cargo aircraft landing at Ellington shortly after Harvey passed through. The Hercules came from the 123rd Airlift Wing of the Kentucky Air National Guard. (DVIDS.)

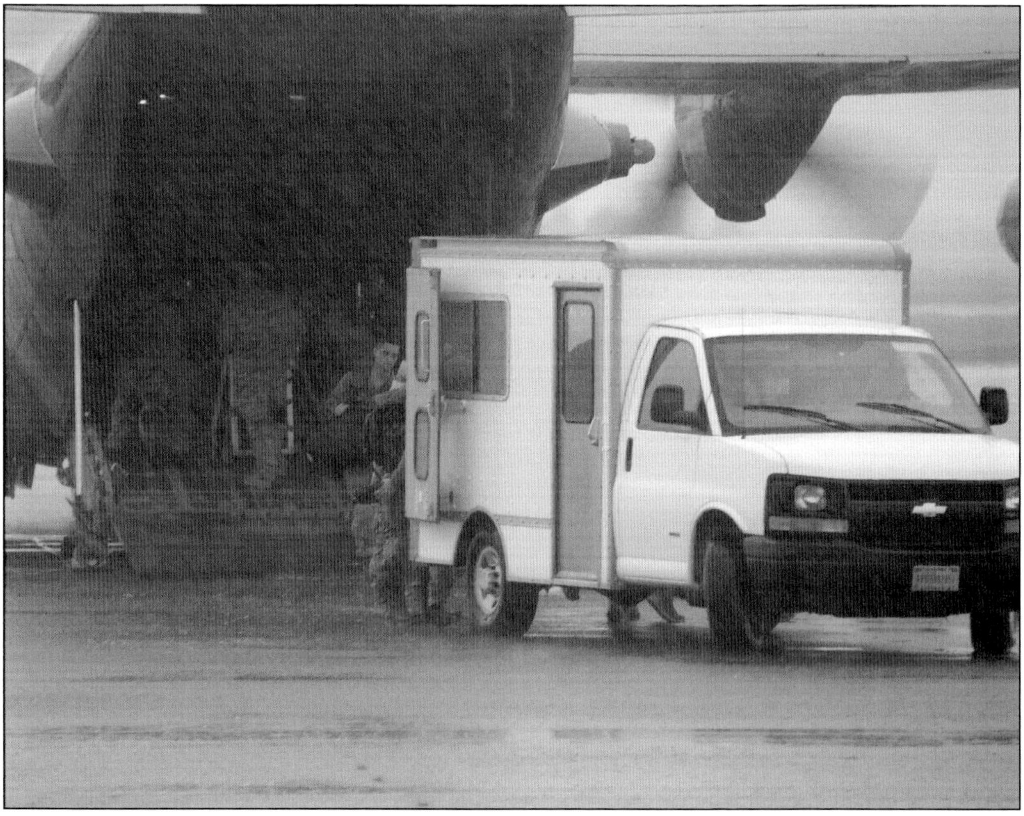

The Hercules carried a Special Tactics Squad to assist with search and recovery operations in Southeast Texas. It is unloading the unit's equipment. (DVIDS.)

Six

The Lone Star Flight Museum 2015–Present

One of the newest additions to Ellington Airport is the Lone Star Flight Museum, which began as a private collection of aircraft. The owner, Robert L. Waltrip, decided to display his collection to the public. In 1990, he created the Lone Star Flight Museum as a nonprofit organization. For the next 18 years, it was located at Scholes Airport in Galveston, Texas. Then, in September 2008, Hurricane Ike came to visit.

The Scholes Airport hangars and the museum's lobby were flooded by seven to eight feet of water when the hurricane passed over Galveston. Most of the displays of aviation memorabilia were destroyed. Static display aircraft, including a B-58 and a Consolidated Catalina (ironically an amphibian aircraft), were severely damaged. Nearly a dozen airplanes of the collection had been flown to safety before the storm, but the museum's damage was too great to permit their immediate return after the storm.

The museum's board decided to move the museum somewhere safer. Ellington Airport had space it was willing to make available. It was only a short distance away, less than 30 miles as the crow (or B-17) flies. It was also 24 feet higher than Scholes. With the aid of corporate donations and a FEMA grant, work started on a new facility at Ellington in November 2015. Work was completed, and the collection moved to the new museum by July 2017, but the opening was postponed until September 16 due to Hurricane Harvey in August 2017. This time, the museum did not flood.

The museum has become an integral part of Ellington since then. It has 25 aircraft in its collection, including civilian and military airplanes and helicopters. It offers rides on seven different aircraft in its flying collection, including its B-17 Flying Fortress and a B-25 Mitchell.

The museum focuses on Texas aviation and is home to the Texas Aviation Hall of Fame. It has also been shaping its collection to highlight Texas contributions to aviation. The museum also has an extensive aviation education program, with an emphasis on STEM (Science, Technology, Engineering, and Mathematics).

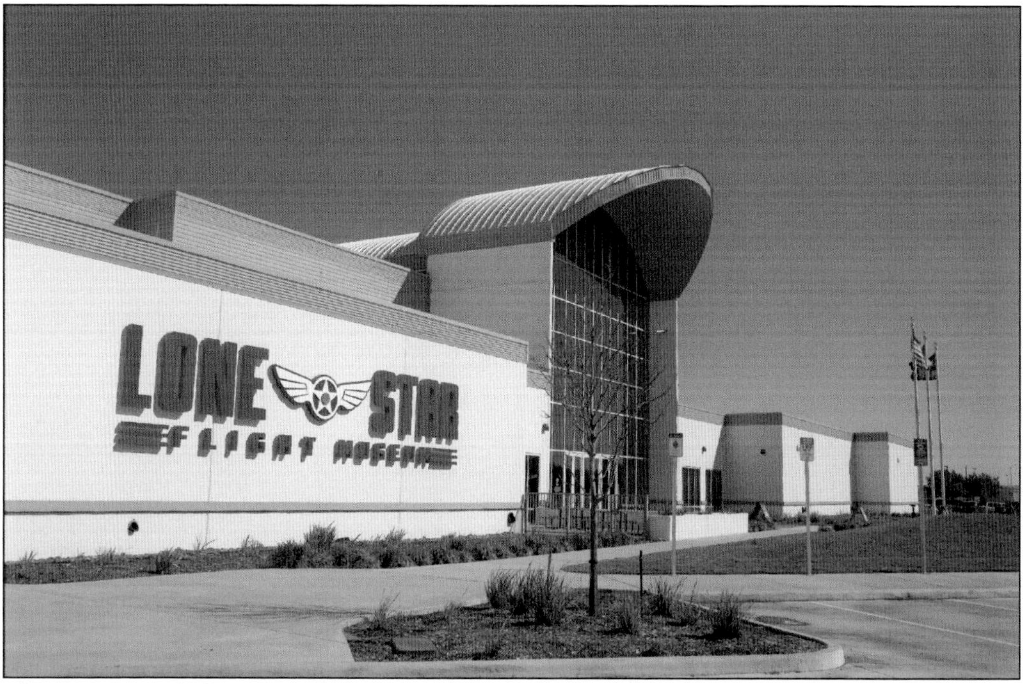

The Lone Star Flight Museum opened its doors to the public at Ellington in 2017. The facility is now the newest major museum in Houston. It is easy to reach. Go through the main entrance to Ellington (Challenger Avenue) and take a left at the KC-135. (WL.)

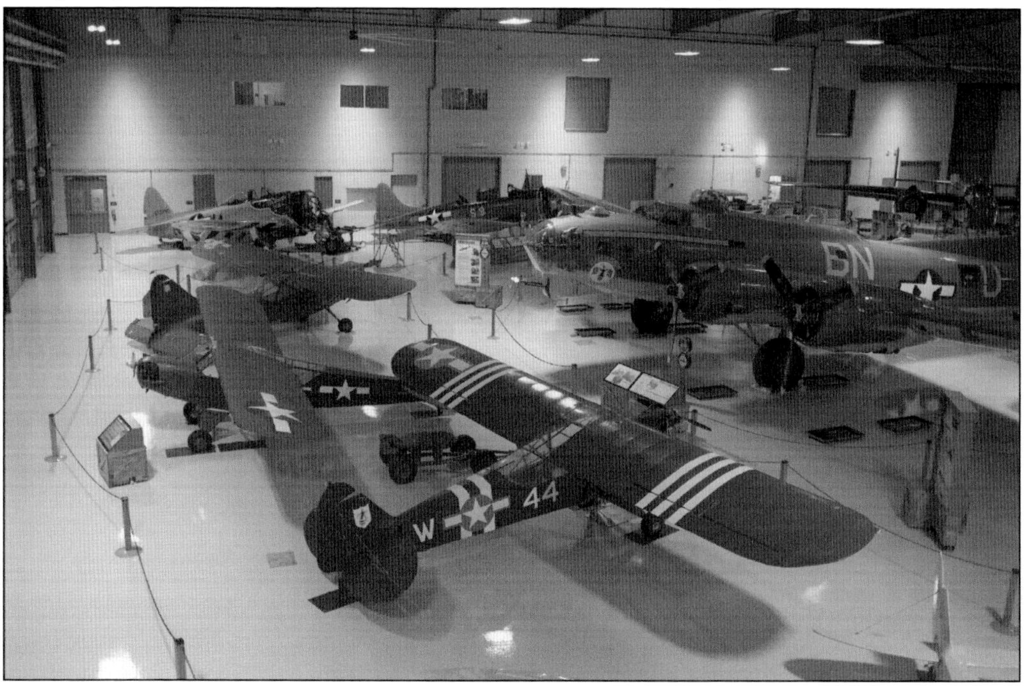

The museum has two galleries filled with aircraft on display. One contains vintage aircraft designed and flown prior to 1945. It holds examples of both military and civilian aircraft of that period. (WL.)

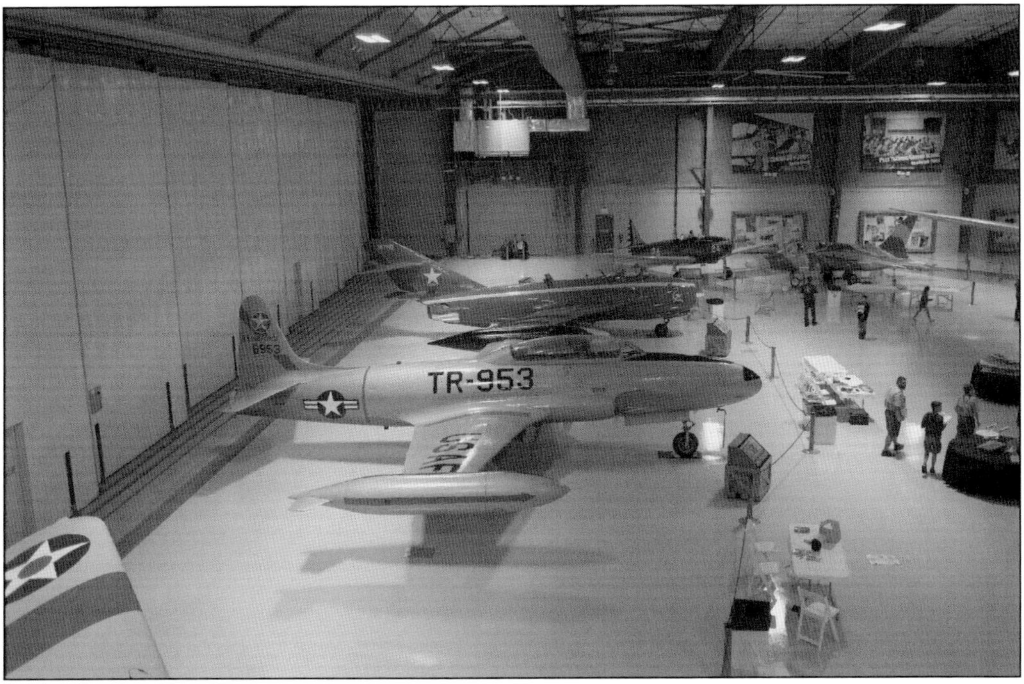

The second gallery is filled with airplanes, jets, and helicopters developed and first flown after 1945. These include warbirds from the United States, Soviet Union, and NATO, as well as modern civilian aircraft. (WL.)

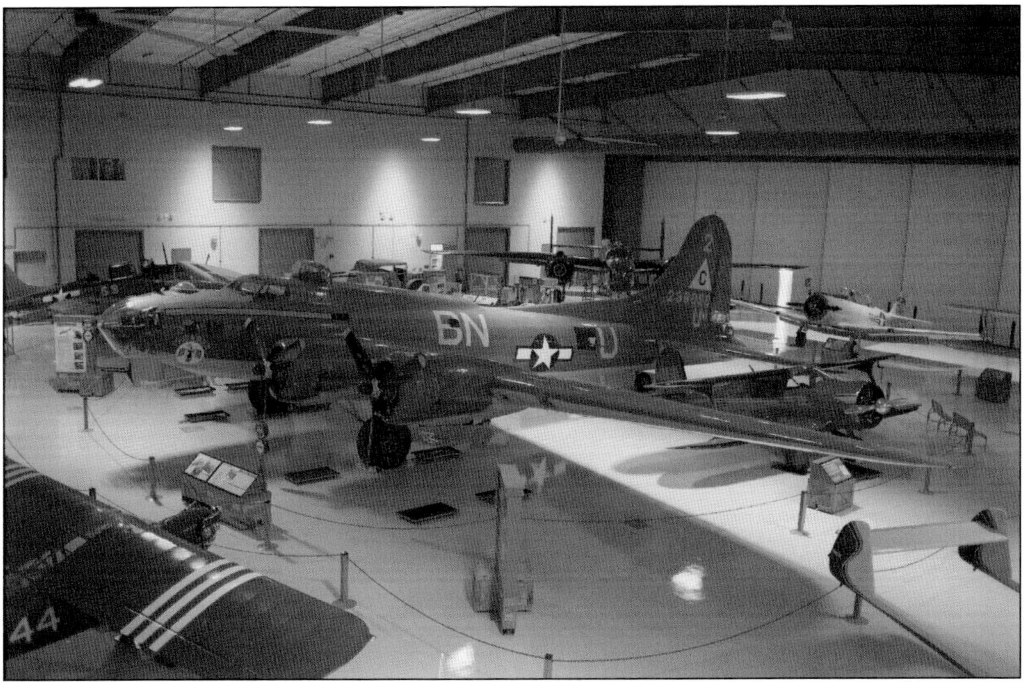

The star of the Lone Star Flight Museum is this B-17G Flying Fortress, *Thunder Bird*. The B-17 formed the backbone of the US air offensive against Nazi Germany. This bomber has been restored to flying condition, and visitors can purchase rides on it. (WL.)

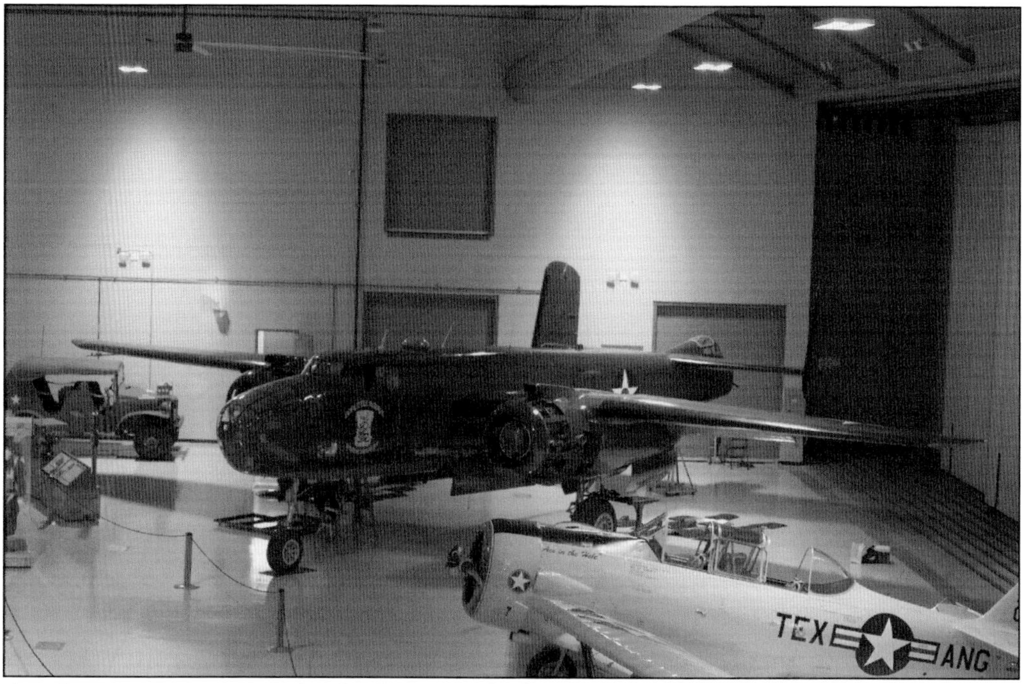

Another star attraction is this B-25. It is marked to appear like one of the aircraft used in the 1942 Doolittle Raid, the first air attack on Japan. It, too, flies and offers rides to the public. This aircraft was donated by Continental Airlines in 2004. (WL.)

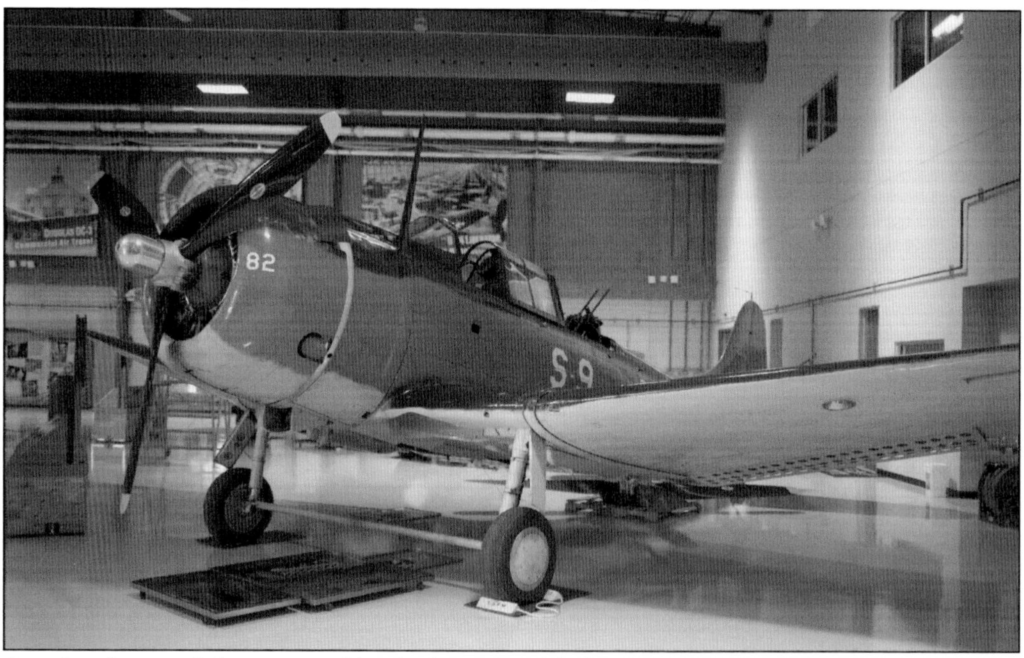

Another flying exhibit possessed by the Lone Star Flight Museum is a Douglas Dauntless. A US Navy dive bomber, the Dauntless played a decisive role at the Battle of Midway, sinking four Japanese aircraft carriers, and continued to play an important role in World War II in the Pacific through 1944. (WL.)

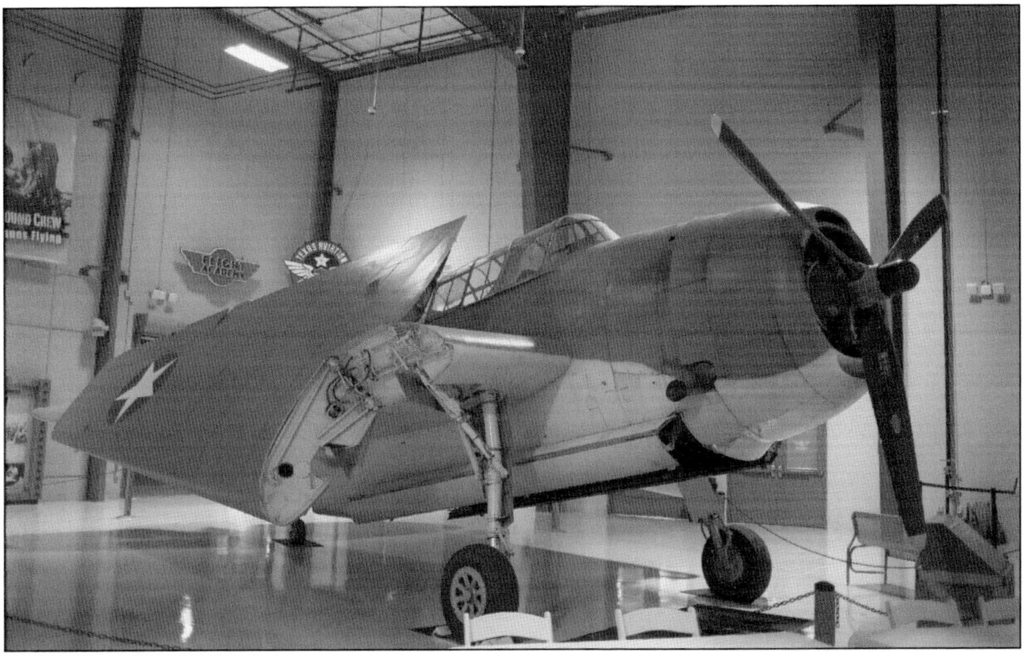

Another important World War II carrier aircraft in the museum's collection is this Grumman TBF Avenger torpedo bomber. In addition to the significant role it played it the Pacific, the Avenger has an important Texas connection. The late Texan and former president George H.W. Bush flew Avengers in combat during World War II. (WL.)

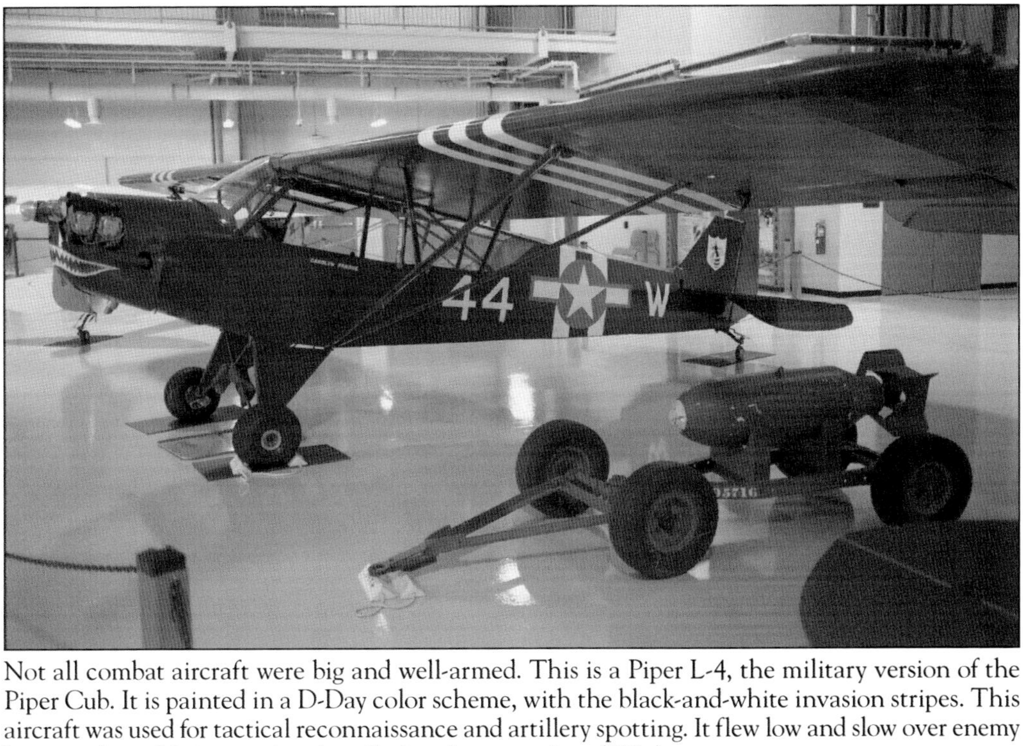

Not all combat aircraft were big and well-armed. This is a Piper L-4, the military version of the Piper Cub. It is painted in a D-Day color scheme, with the black-and-white invasion stripes. This aircraft was used for tactical reconnaissance and artillery spotting. It flew low and slow over enemy lines, with its fabric exterior the pilot's only protection. (WL.)

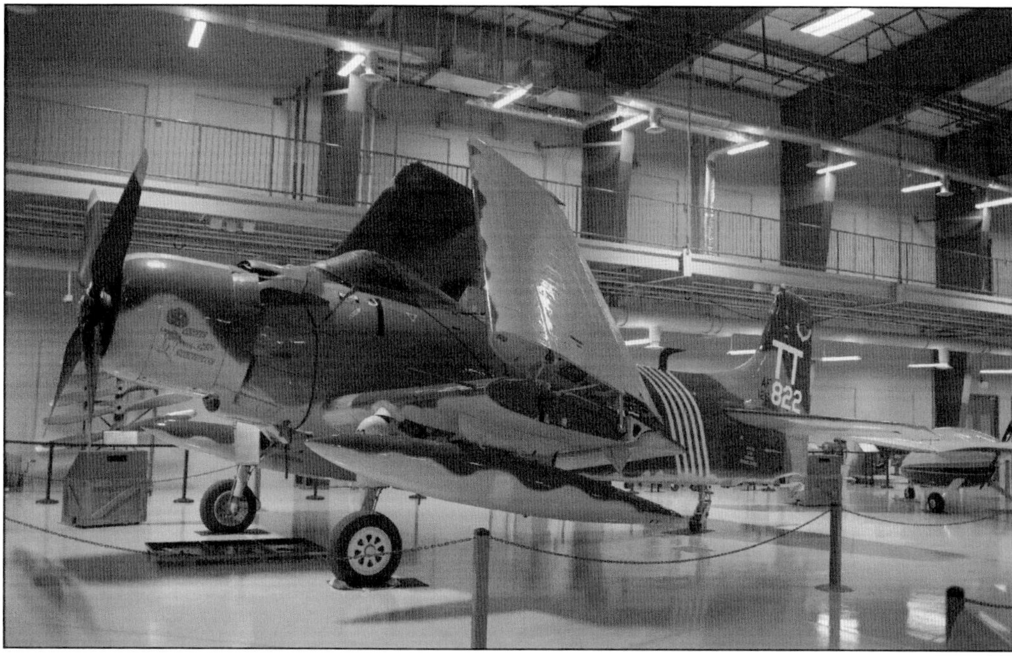

The Lone Star Flight Museum also has aircraft that flew in the 1950s and during the Vietnam War. This is a Douglas Skyraider, an airplane used to provide air support to US and ARVN troops in Vietnam. It was designed to fly off aircraft carriers, but its short-field performance was useful for operating on short land-based airfields. (WL.)

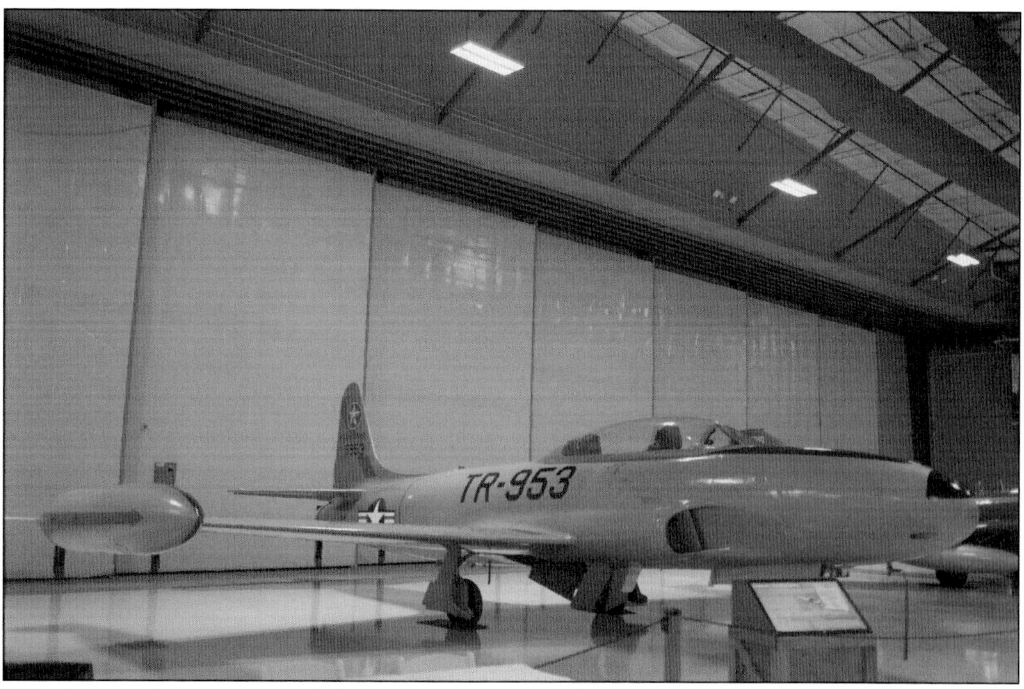

Whether as the single-seat fighter version F-80 or the twin seat trainer T-33, the Lockheed Shooting Star played an important role at both Ellington and the Air Force squadrons assigned there. The Lone Star Flight Museum has a T-33A in Ace-in-the-Hole squadron markings. (WL.)

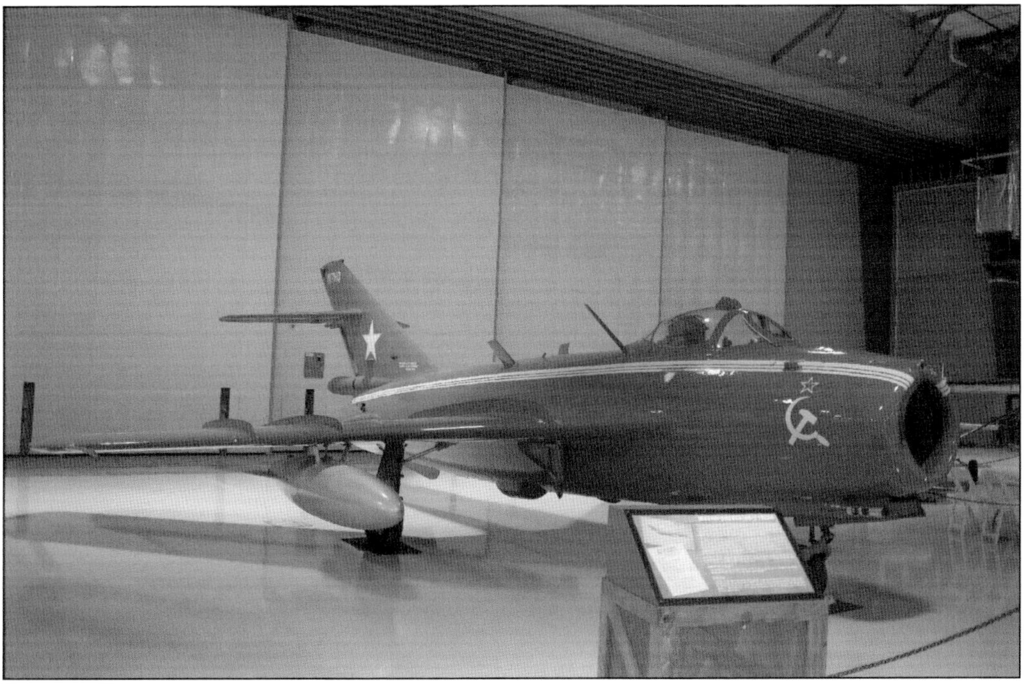

The museum also has a mint-condition MiG-17, *Fresco*, painted in Soviet red and yellow. This aircraft was one of the 111th Squadron's most dangerous foes when the Ace-in-the-Hole fought in the Korean War. It is not a big aircraft, but in its day it was deadly. (WL.)

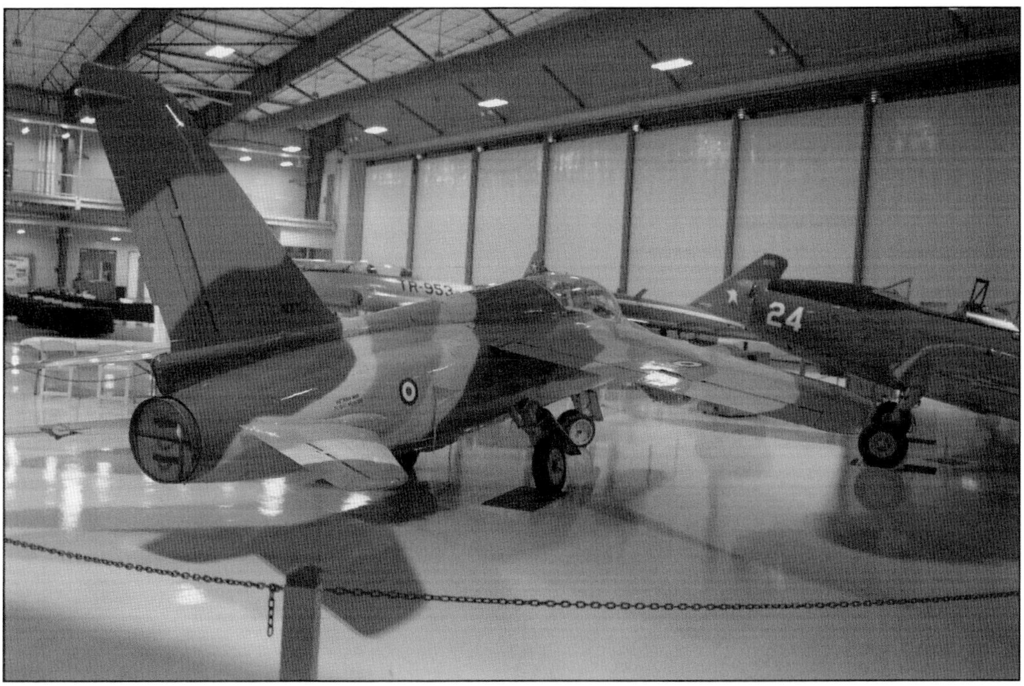

The museum also has a Folland Gnat, an aircraft designed as an affordable subsonic fighter in Great Britain. It first flew in 1955. While the RAF never used it as a fighter, Finland and India both did. This aircraft shows the international scope of the Lone Star Flight Museum's collection. (WL.)

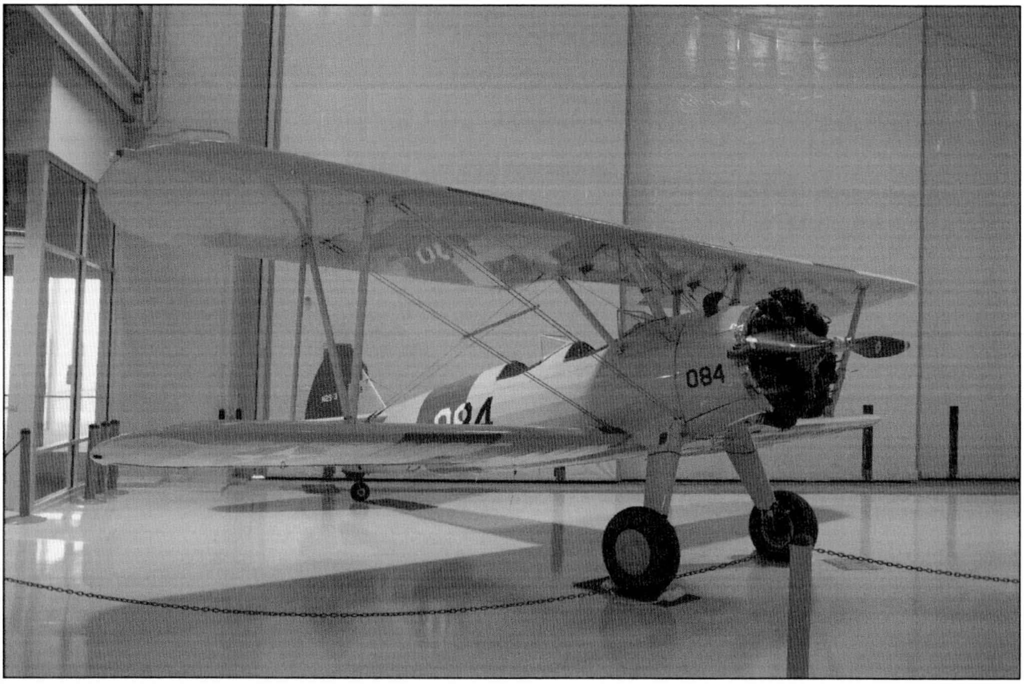

There are some classic military training aircraft at the museum, including this Boeing-Stearman PT-17 Kaydet. First flown in 1934, it was used as a primary trainer by both the US Navy and the Army Air Force during World War II. The museum's Kaydet is in flyable condition. (WL.)

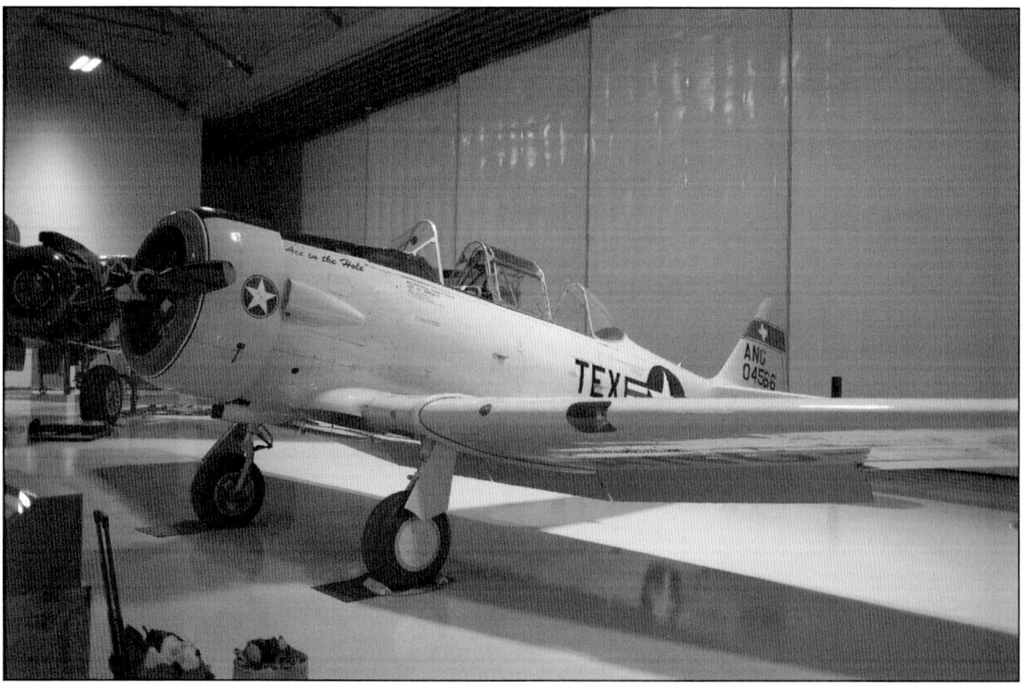

The T-6 Texan (also known as the SNJ when in US Navy service) was an advanced trainer during World War II. The museum's T-6 flies, and is available for rides by visitors. (WL.)

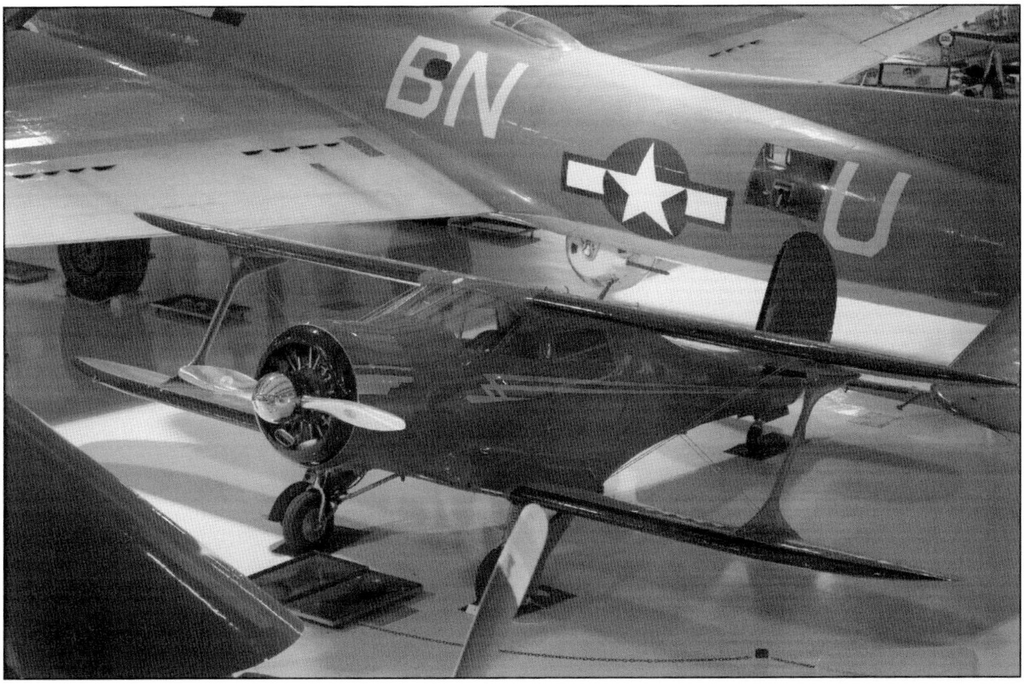

The collection is not all warbirds. The Beech Staggerwing was *the* executive aircraft of the 1930s and 1940s. Despite being a biplane, it could cruise at 202 mph, with the pilot and three passengers in an enclosed cockpit. This is one of only 785 built and 10 surviving Staggerwings. (WL.)

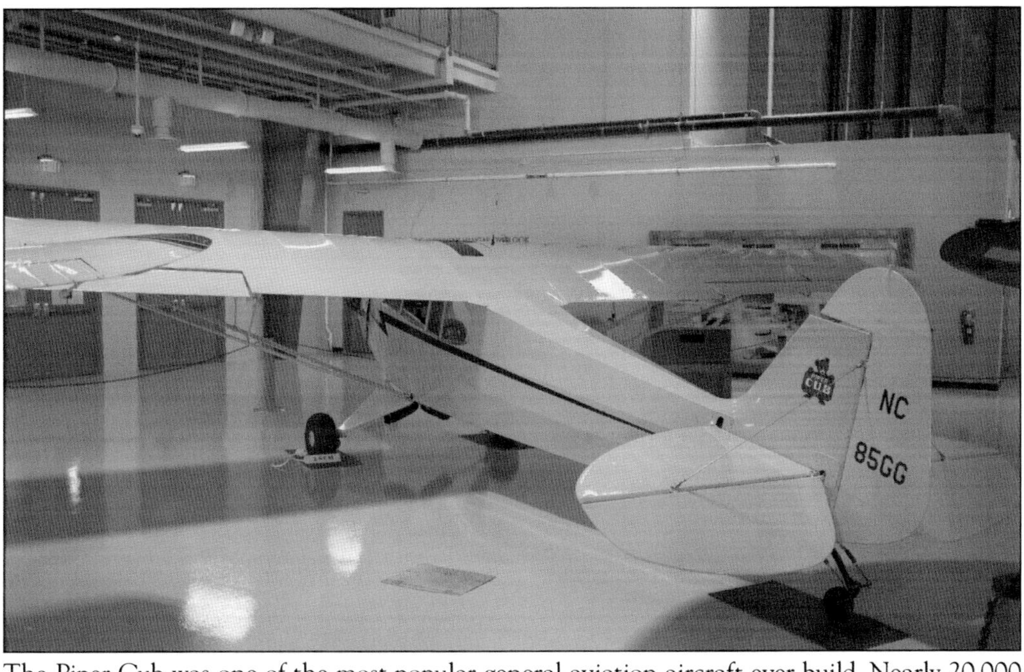

The Piper Cub was one of the most popular general aviation aircraft ever build. Nearly 20,000 were built. The museum has a military version of the Cub in its collection, but it also has this civilian Cub painted in Piper Cub yellow. A generation of general aviation pilots learned to fly on the Cub. (WL.)

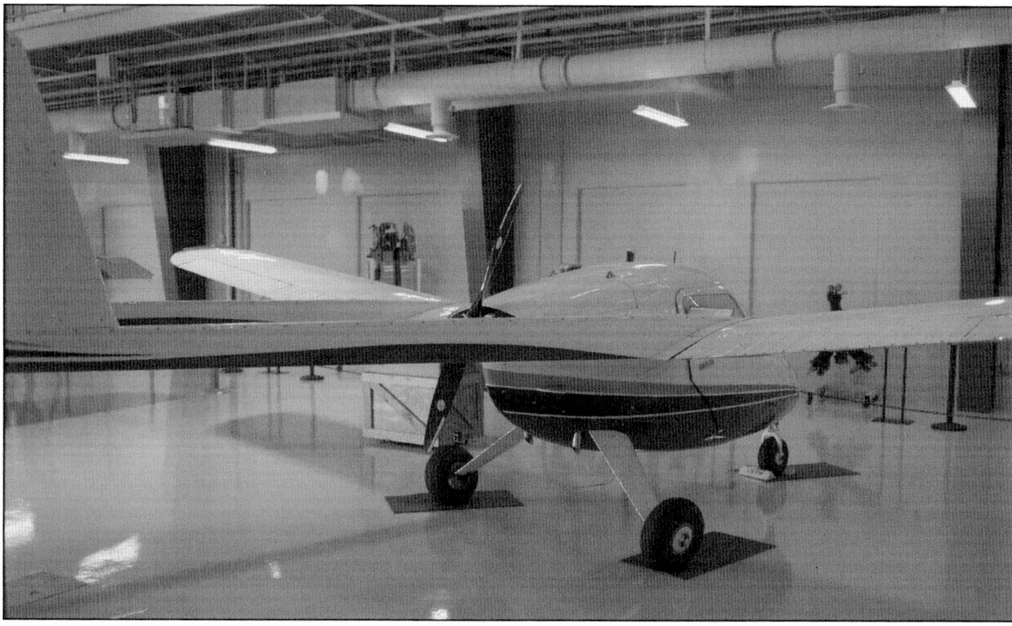

Among the rarer birds in the Lone Star collection is this Anderson-Greenwood AG-14. Three Texas entrepreneurs designed it for the general aviation market in early 1950. Unfortunately, the Korean War kept it from going into production, and only five were built. The museum got one AG-14 as part of its mission highlighting Texas aviation. The aircraft was to have been built in Houston. (WL.)

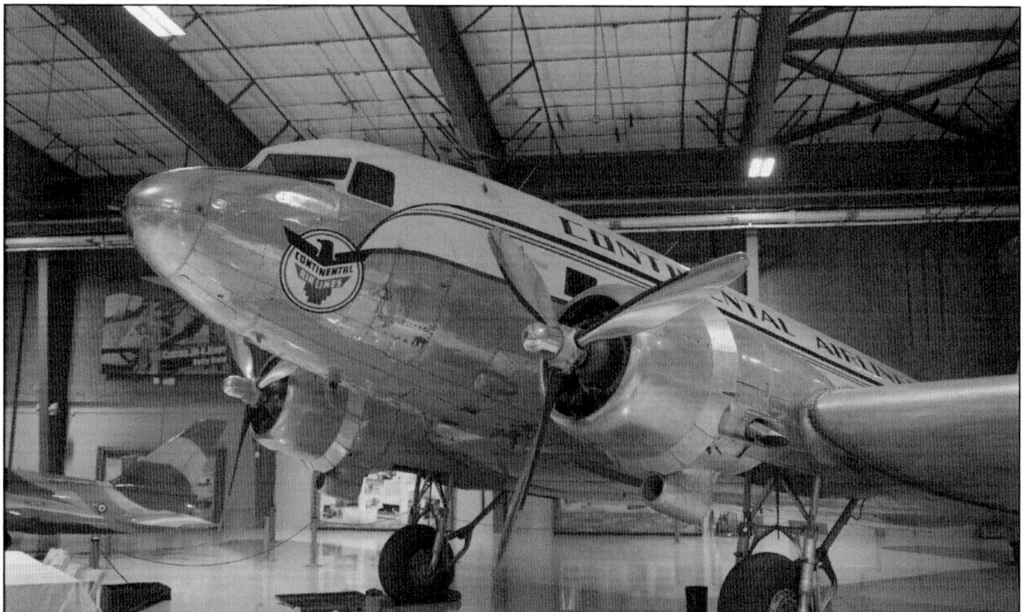

Another aircraft on display at the Lone Star Flight Museum is this DC-3. The DC-3 was the most successful airliner built before the introduction of jet airliners. The museum's DC-3 is painted in the livery of Continental Airlines, which was headquartered in Houston prior to its acquisition by United Airlines. (WL.)

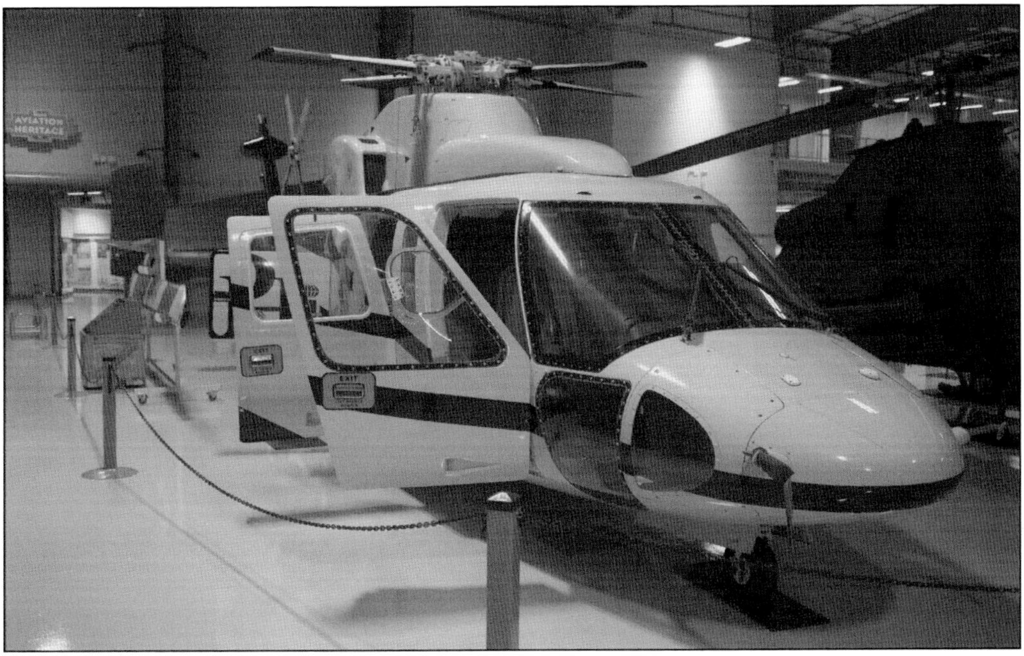

The offshore oil industry is a big part of Texas's economy. Helicopters like this Sikorsky S-76A transported supplies and crew to and from oil platforms. This helicopter was designed to support the oil industry, and the museum's example served Petroleum Helicopter Inc. for 33 years, making over 36,000 takeoffs and landings in its career. (WL.)

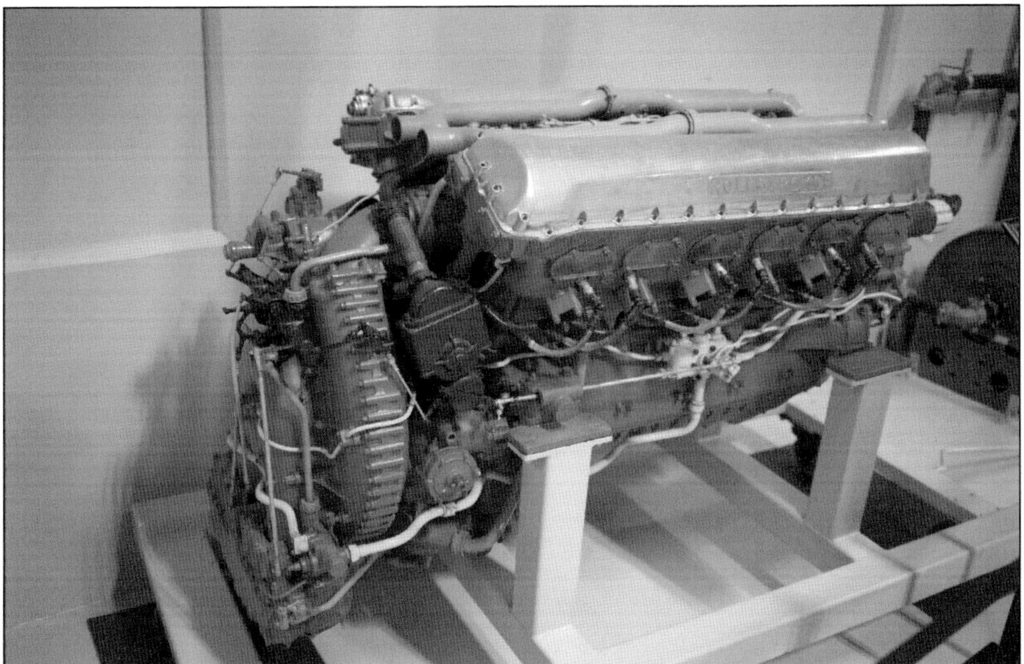

The Lone Star Flight Museum also has collections of aviation-related artifacts. This includes a fine collection of aircraft engines. One of them is this Rolls-Royce Merlin engine. It powered the fabled Hurricane, Spitfire, and Mustang fighters as well as the Lancaster bomber. (WL.)

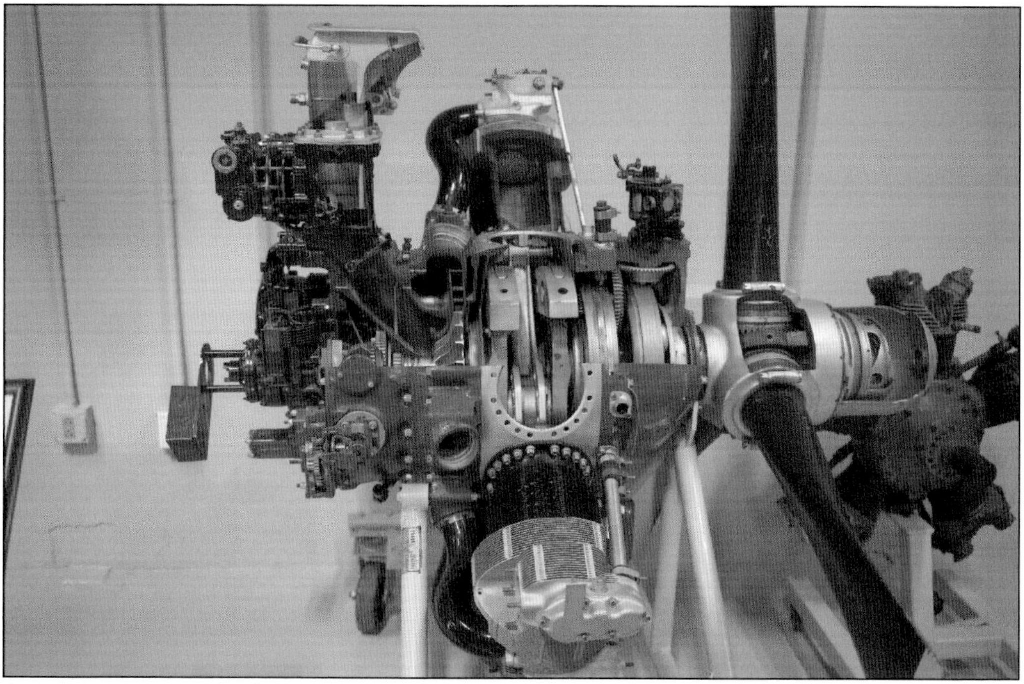

Another engine on display is the Wright R-1820 Cyclone. An air-cooled radial engine, it was used on aircraft as varied as the B-17, SDB Dauntless, and DC-3. The displayed engine is cut away to display its inner workings. (WL.)

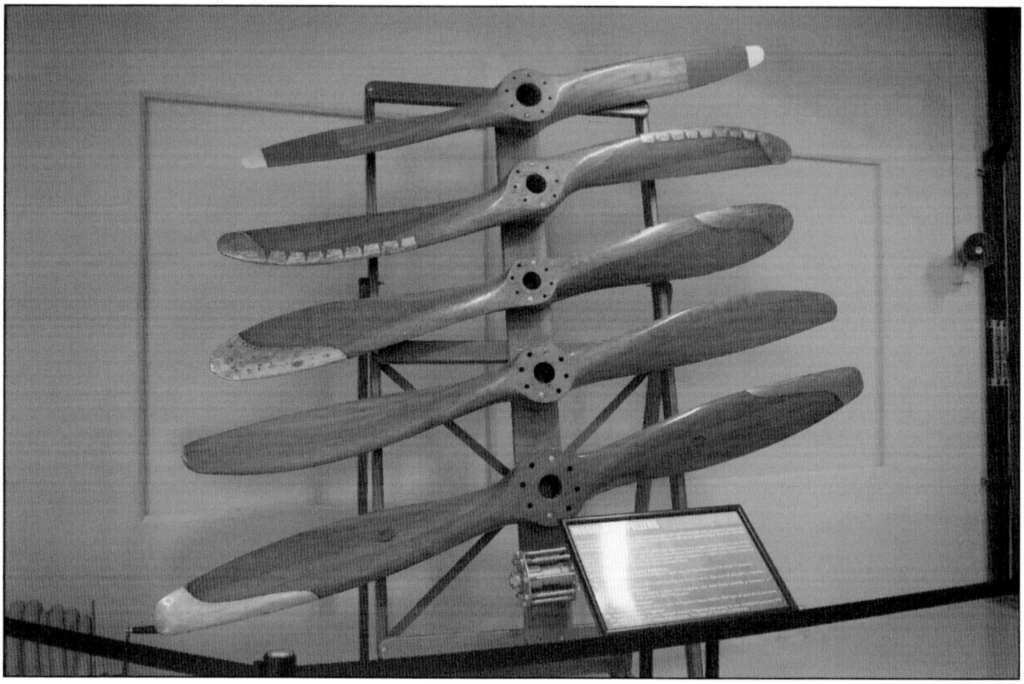

There is also a display of 1930s and 1940s twin-bladed wooden propellers. State-of-the-art at the start of the 1930s, they were replaced by variable-pitch metal propellers for high-performance aircraft by the 1940s, but continued to be used in general aviation well into the 1960s. (WL.)

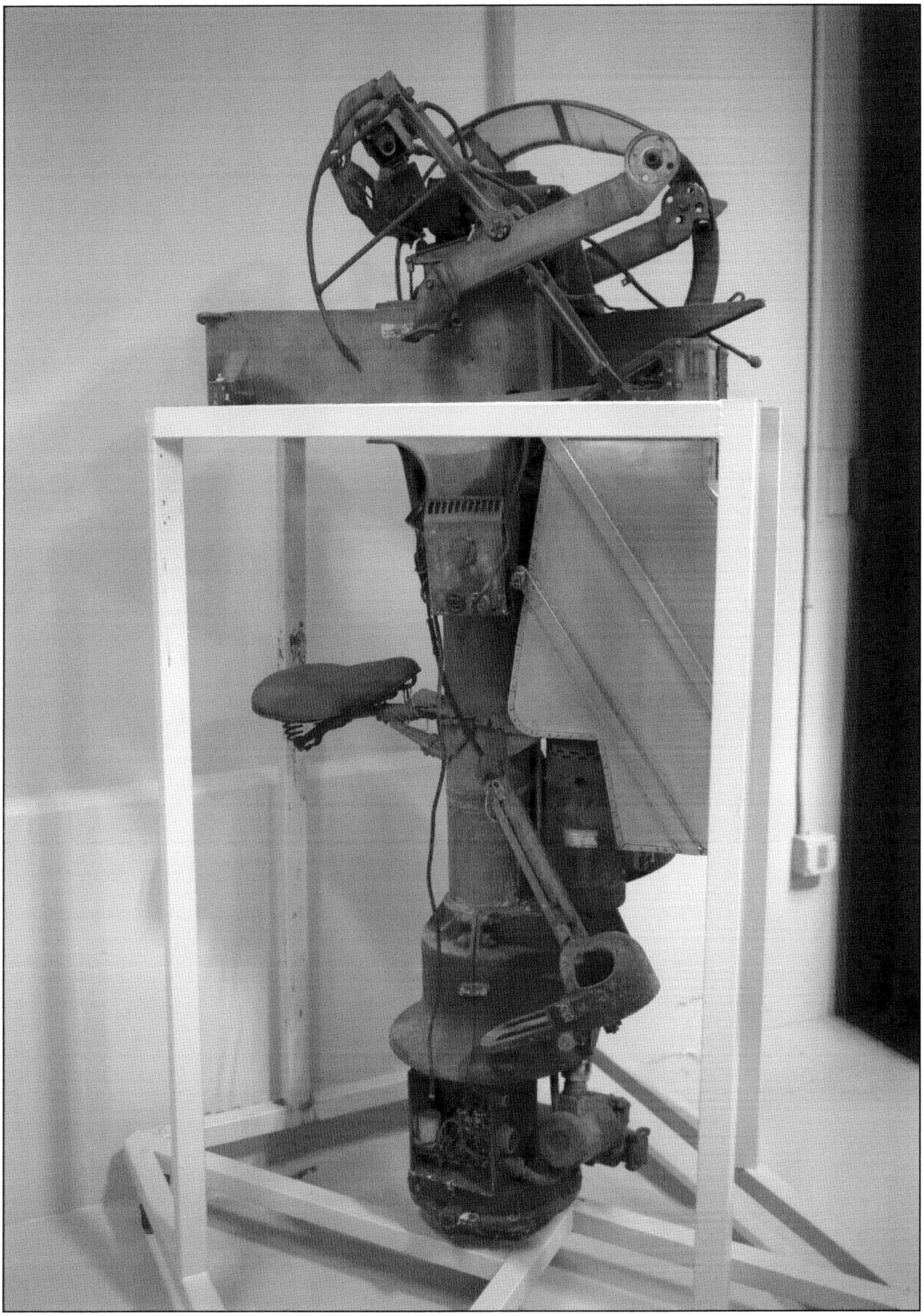

Another artifact on display is this Bendix Model L mid-upper turret. This example came off a B-25 Mitchell, but similar turrets were used on the B-26, B-17, and B-24 bombers. This turret was electric powered; the motor turned the turret to the gunner's commands. (WL.)

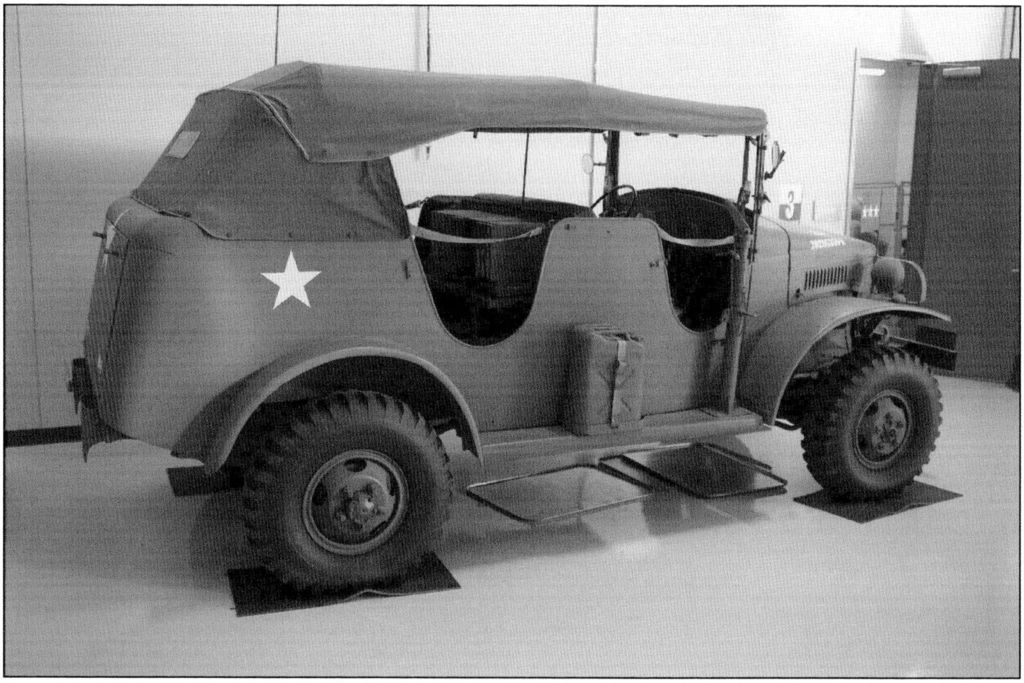

The museum also has this 1940 Dodge half-ton staff car. It is flying the flag of a US Army (or more likely Army Air Force) lieutenant general. (WL.)

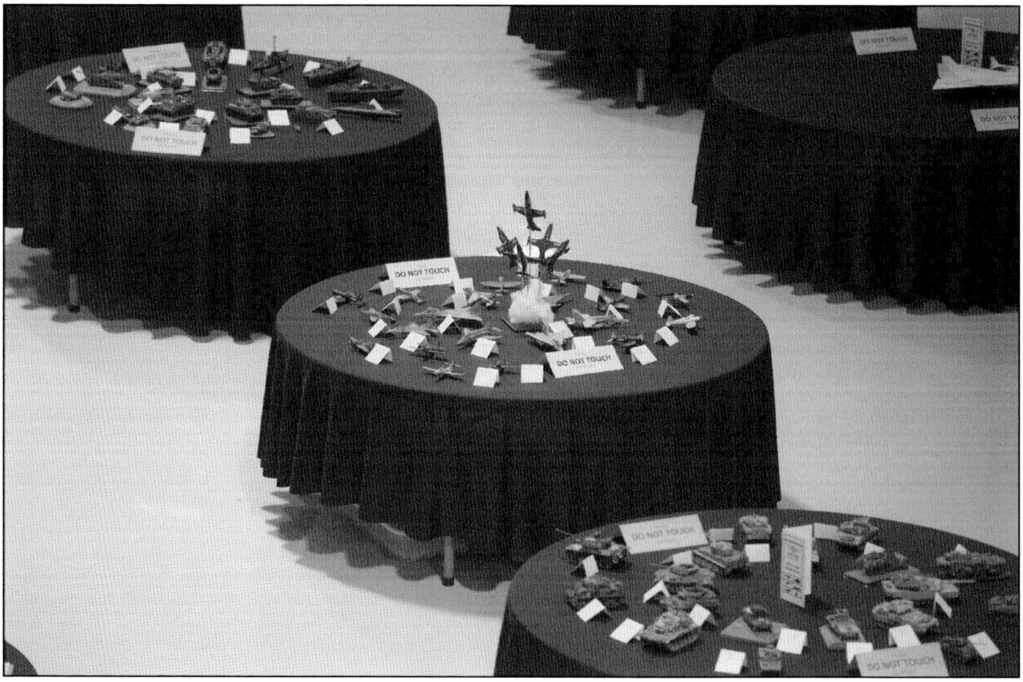

The museum is also the venue for special events and shows. In January, it hosts its annual model expo. Modelers from Harris, Galveston, and surrounding counties bring models to display to the public and each other. The day also features model-making workshops. (WL.)

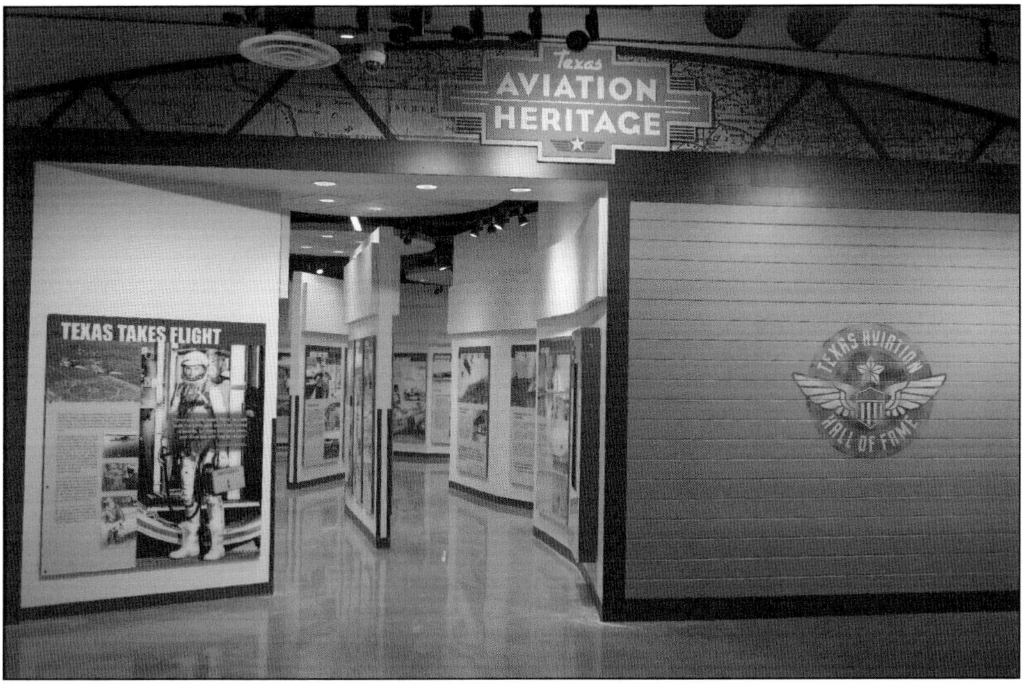

The museum also has the Texas Aviation Heritage Gallery. Visitors can take a walk through the aviation history of Texas, from flying's earliest days prior to World War I to the 21st century. (WL.)

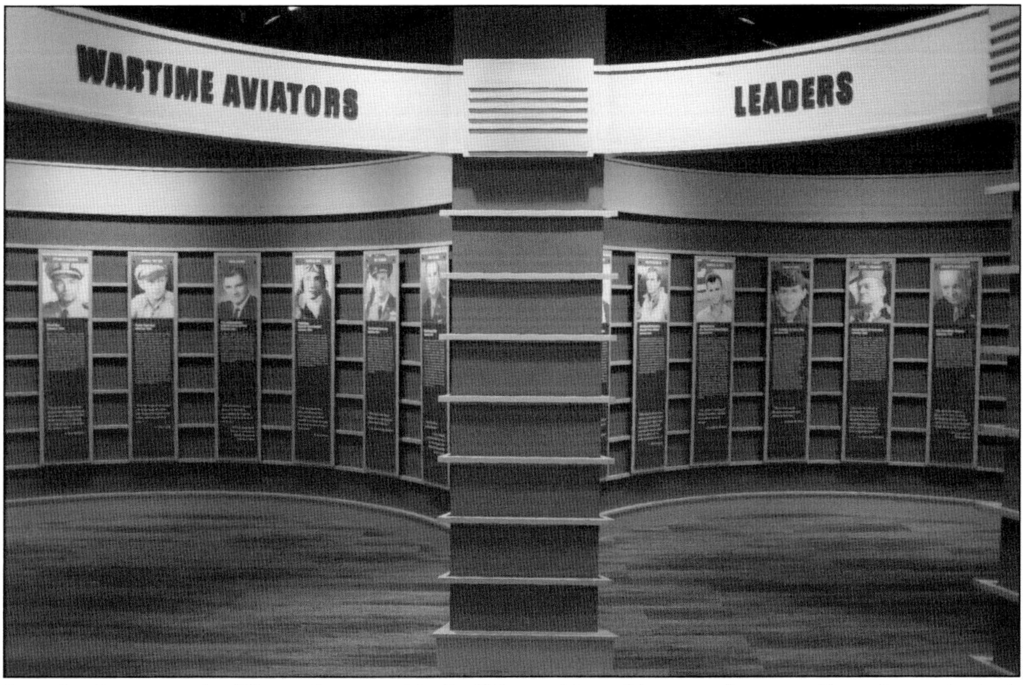

It is also home to the Texas Aviation Hall of Fame. Here, Texans who contributed to aviation's history are noted and recognized. New members are inducted annually. The inductees include aircraft manufacturers, pioneer fliers, wartime heroes, business leaders, and aerospace pioneers. (WL.)

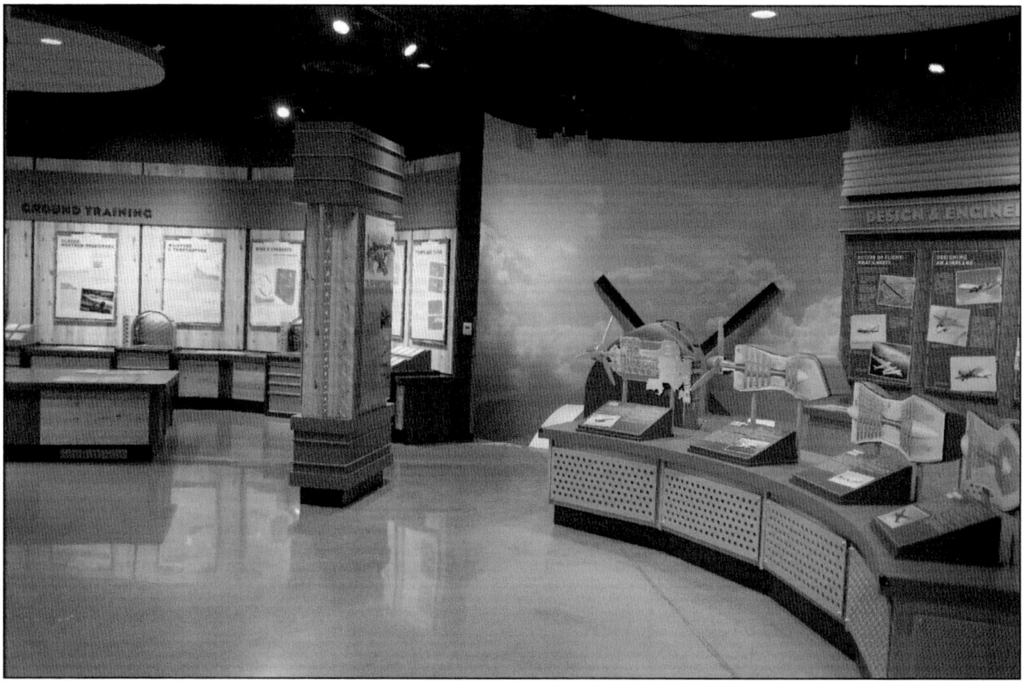

Education is also a major mission for the Lone Star Flight Museum. This includes the Learning Lab extensive displays on how aircraft work and are designed, along with navigation, weather, and other concepts critical to successful flight. (WL.)

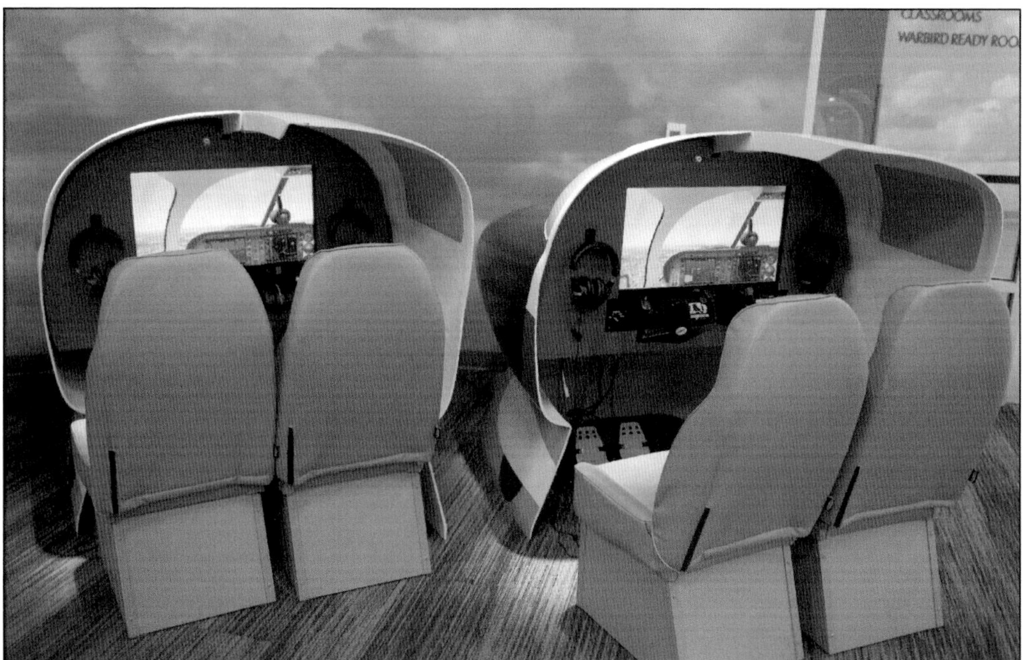

The museum also has the Flight Simulator Bay, where students can test their skill in a simulator of a Mooney Ovation light aircraft. Students can use the simulators to "fly" from Ellington Field to Scholes Field on Galveston Island. (WL.)

Seven

To Tomorrow and Beyond

The Future of Ellington

Today, Ellington Airport is many things. The Houston City Council renamed it Ellington International Airport in 2013, although the FAA still calls it Ellington Airport. It remains the home of the Ellington Field Joint Reserve Base, Coast Guard Air Station Houston, and Johnson Space Center's aircraft fleet. The Lone Star Flight Museum is expected to draw visitors to Ellington. At the end of its first year of operation at Ellington, nearly 50,000 individuals were admitted from 47 states and 18 countries.

Although there are no scheduled passenger or air freight operations out of Ellington, it remains important for Houston's general aviation community. It is the one municipal airport in Houston that does not impinge on commercial aircraft operations.

Further, the City of Houston has conducted a major upgrade of Ellington's facilities in the last 10 years. This includes construction of general aviation hangars, creation of a business incubator, and the recent completion of a state-of-the-art 21st century control tower.

One important and exciting 21st century development at Ellington is the creation of Spaceport Houston at Ellington. In 2014, Sierra Nevada Corporation signed an agreement with the Houston Airport System to explore developing Ellington into a commercial spaceport. Small spacecraft could be launched into space from Ellington's runways using commercial space planes, such as Sierra Nevada's Dream Chaser.

Houston Airport System sought and received a launch site license from the FAA for Ellington in June 2015, and Ellington became the 10th commercial spaceport in the United States. In June 2019, ground was broken on the Houston Spaceport at Ellington Airport. Planned is a 154-acre phase 1 development to support spacecraft operations from Ellington. It is the beginning of a 450-acre spaceport planned on the south side of Ellington, near the existing Sonny Carter facility. It will support commercial space travel, with subsonic, supersonic, and hypersonic spacecraft and aircraft operations.

When the 20th century opened, Ellington Field pioneered aircraft flight. A century later, moving into the 21st century, it promises to pioneer commercial spaceflight. Eric Ellington would be amazed.

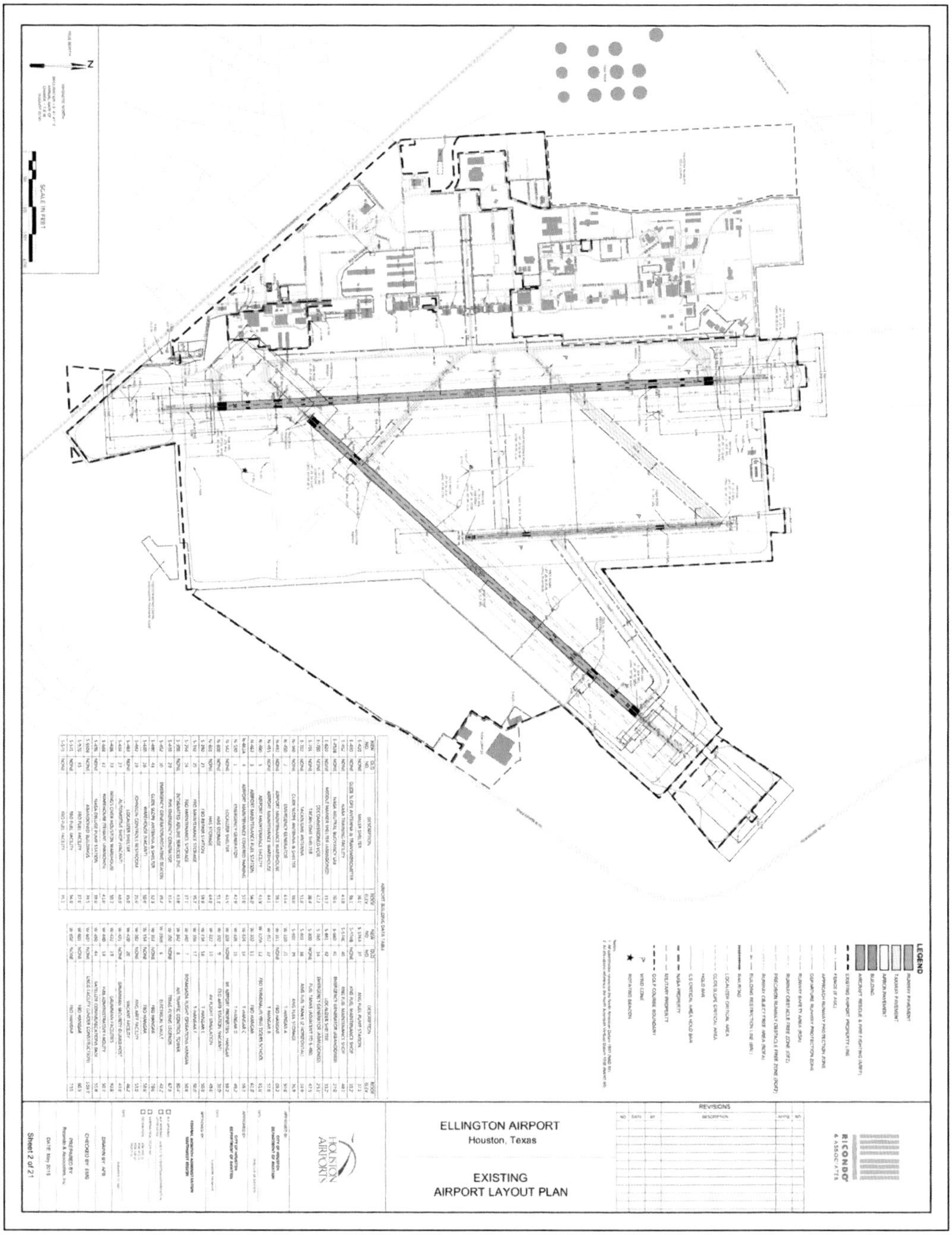

The Ellington Airport layout plan as of 2015 is shown here. The Lone Star Flight Museum is not included. Note the largest remaining undeveloped space is on Ellington's southeast side, paralleling Space Center Boulevard. This is where the Houston Airport System intends to place the Houston Spaceport. (HAS.)

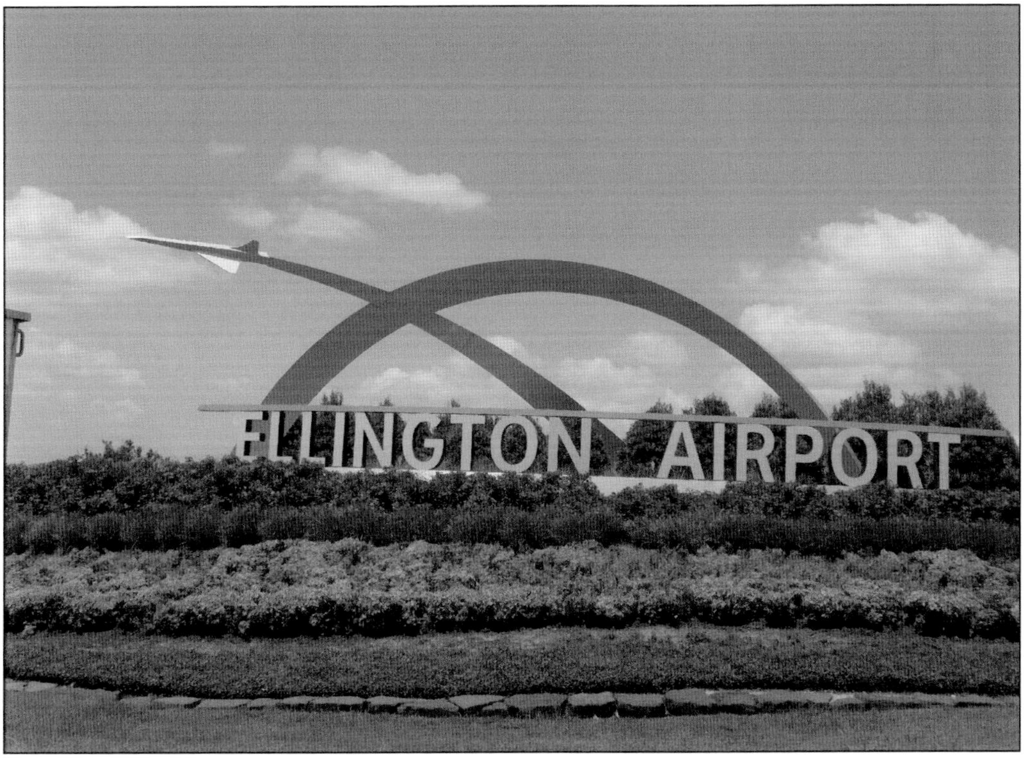

This sculpture-sign greets visitors entering Ellington Airport through the main entrance, off State Highway 3. (AP.)

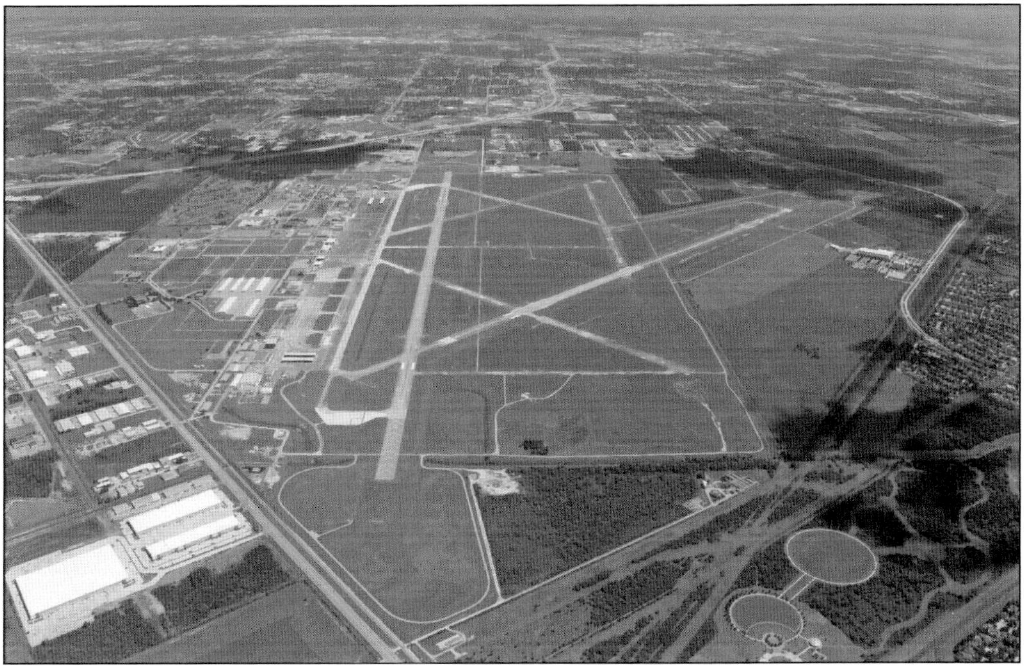

This photograph shows Ellington Airport looking north from a cockpit along Ellington's main runway, 5 Right. (HAS.)

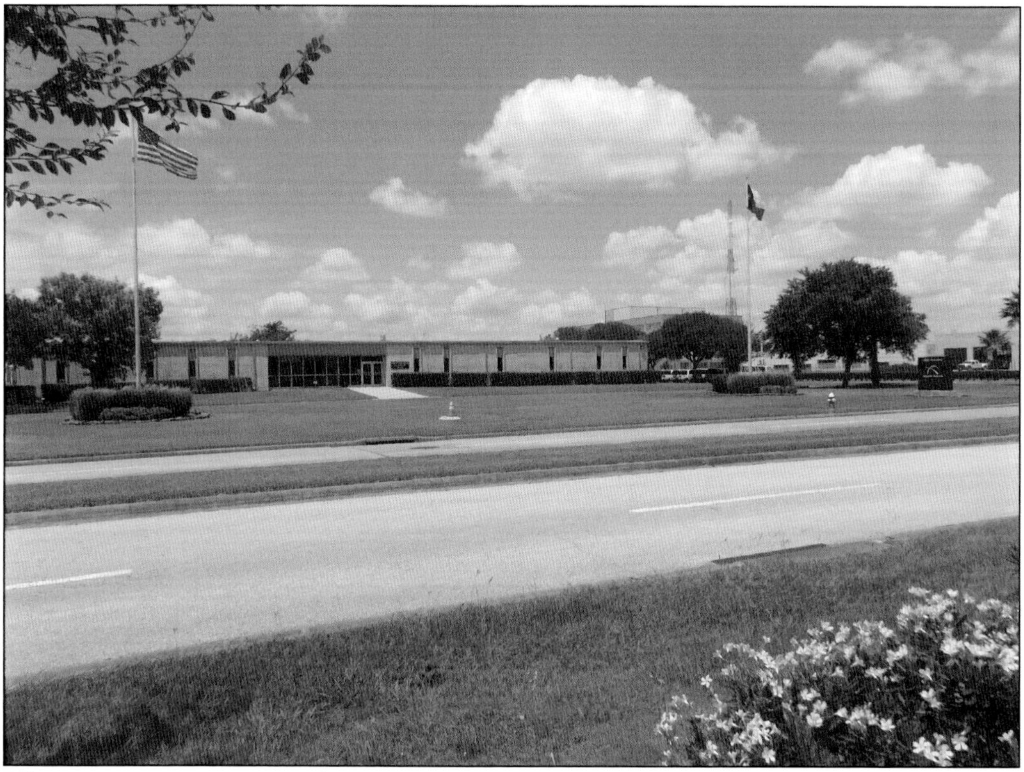

Ellington Airport's current administrative building is seen here. It was once the post hospital, built during Ellington's days as an Air Force base during the Cold War. (AP.)

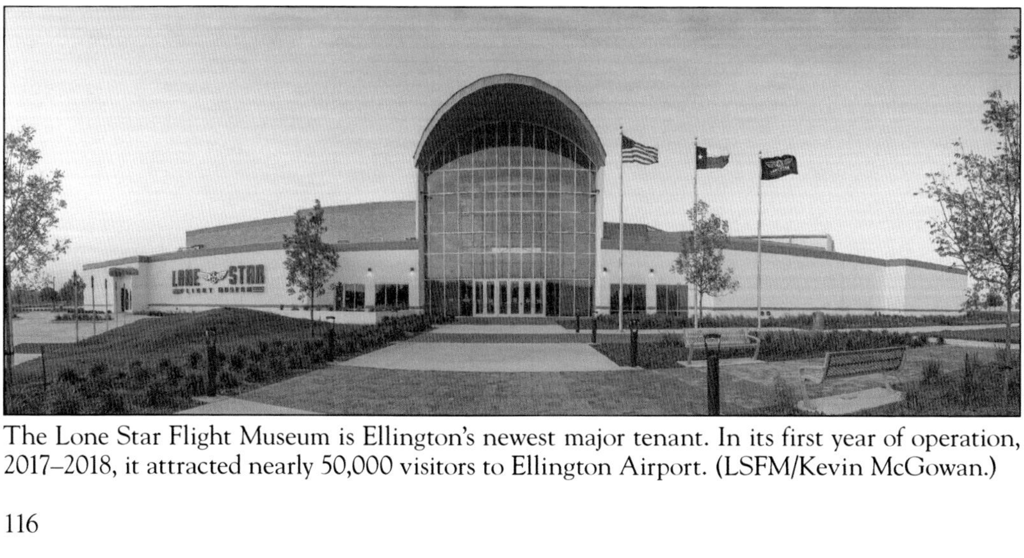

The Lone Star Flight Museum is Ellington's newest major tenant. In its first year of operation, 2017–2018, it attracted nearly 50,000 visitors to Ellington Airport. (LSFM/Kevin McGowan.)

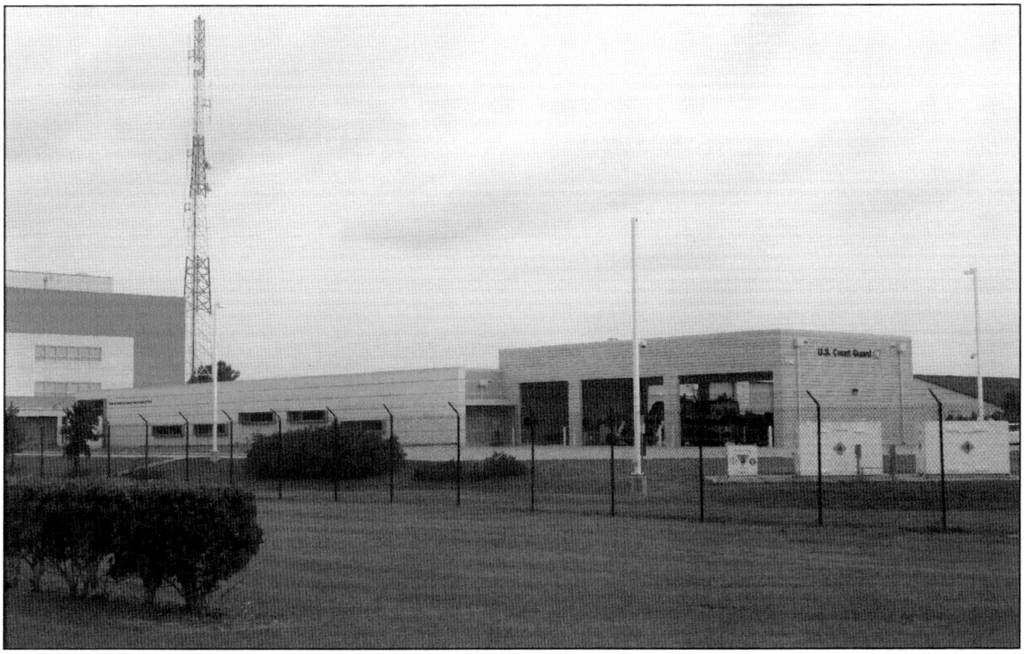

Also on Aerospace Boulevard, next door to the Administrative Building, is the new US Coast Guard building. From here, the Coast Guard provides a protective overwatch of the Houston Ship Channel, Galveston Bay, and the Gulf of Mexico near Galveston Island and the Bolivar Peninsula. (AP.)

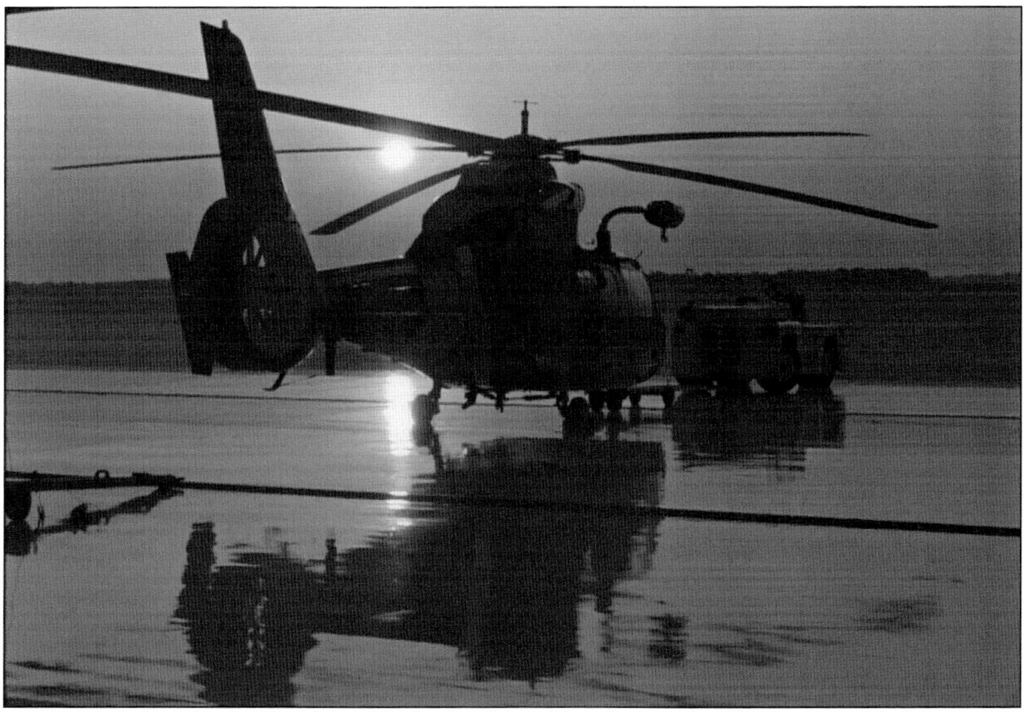

The Coast Guard continues to operate MH-65 Dauphin helicopters out of Air Station Houston. Currently, there are eight MH-65s stationed at Ellington. (DVIDS.)

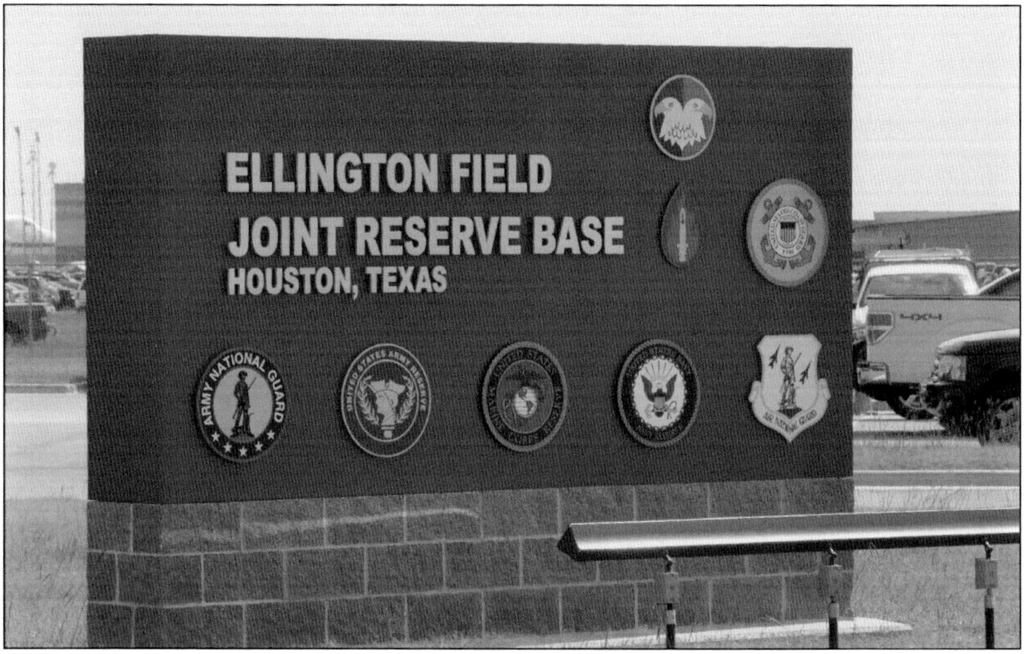

Ellington Airport also continues to host the Ellington Field Joint Reserve Base. It is truly a joint facility. Components of the Army National Guard, the Army Reserve, the Marine Corps Reserve, the Navy Reserve, the Air National Guard, and the Coast Guard all have units that call Ellington Joint Reserve Base home. (DVIDS.)

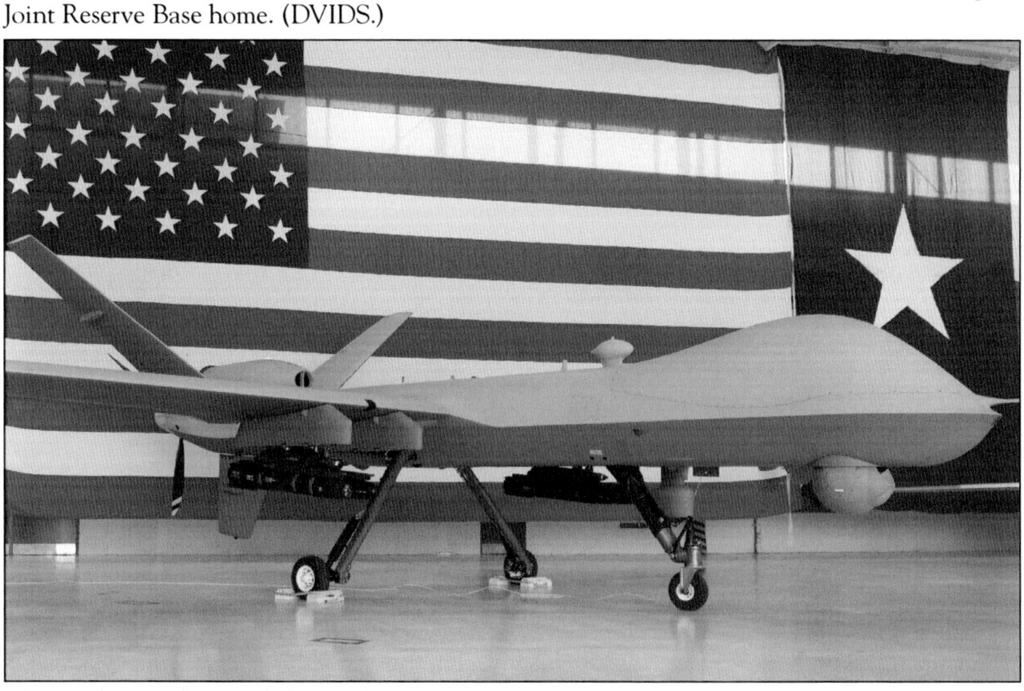

The 111th Squadron and the 147th Attack Wing, historically part of Ellington since the end of World War II, still call Ellington home. Instead of flying F-16s, they now operate drones. In 2017, they converted from the MQ-1B Predator to this aircraft, the MQ-9 Reaper. Pilots based at Ellington can attack targets on the other side of the globe, including Afghanistan or Syria. (DVIDS.)

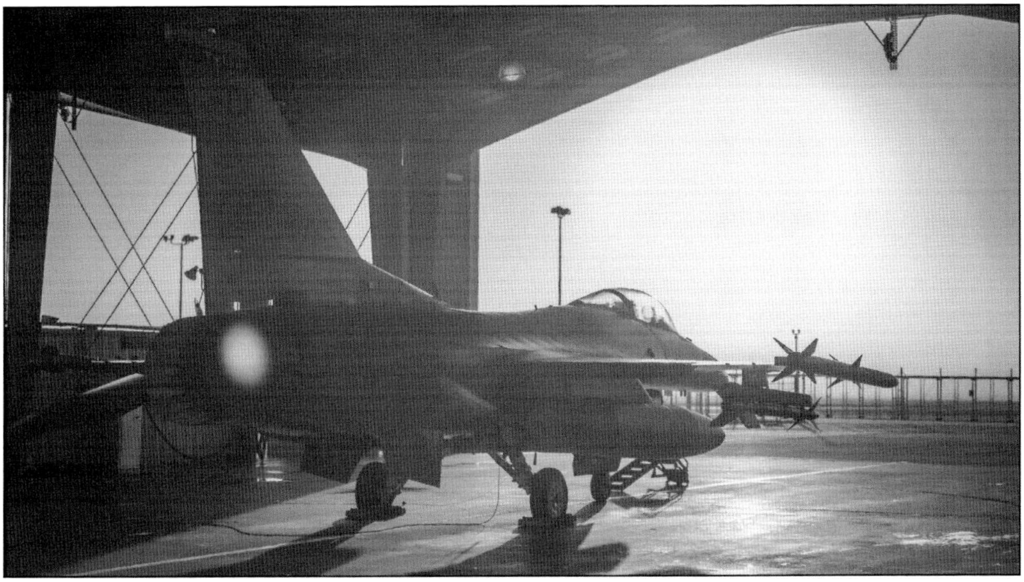

Facilities to operate F-16s still exist at Ellington, and the Fighting Falcon can be staged out of Ellington when the need arises. This Air National Guard F-16 Fighting Falcon from the 138th Fighter Wing Detachment 1, normally part of the Oklahoma Air National Guard stationed at Tulsa, operated out of Ellington in 2017 as part of an exercise. (DVIDS.)

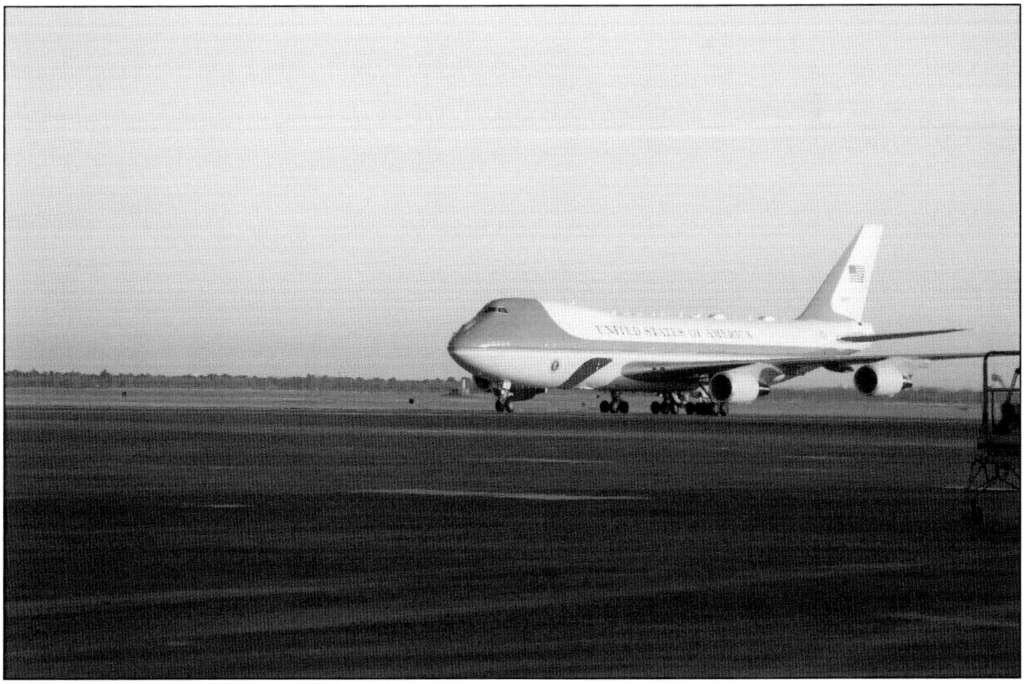

Ellington continues to be the stop for Houston-bound VIPs. Air Force One remains a frequent visitor. (DVIDS.)

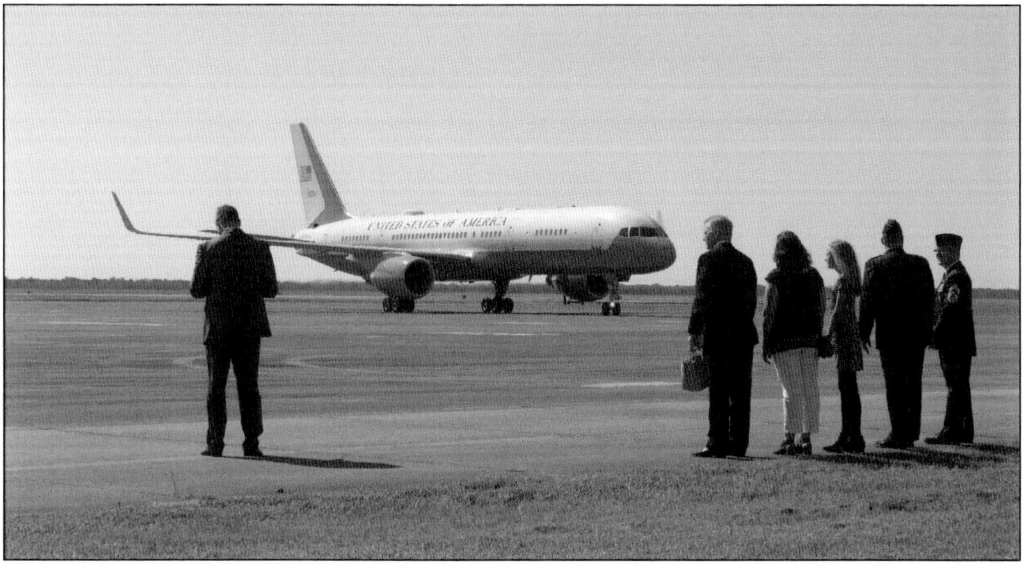

Another frequent visitor is the aircraft used most often as Air Force Two, a Boeing VC-32, a 757 variant modified to carry important passengers. It is most often used to carry the vice president, and is shown here after landing at Ellington. (DVIDS.)

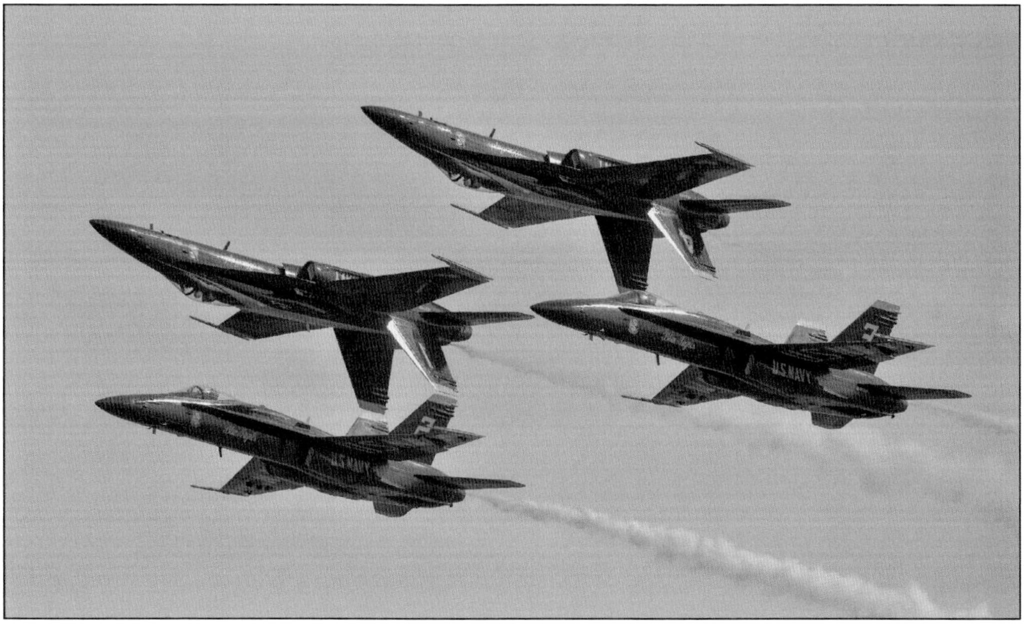

Wings over Houston continues to remain a part of Ellington, and the Blue Angels remain frequent visitors. Today, however, they fly the FA-18 Hornet, one of the hottest aircraft in the Navy's inventory. The routines remain familiar, however. (DVIDS.)

HAS is upgrading Ellington Airport for the 21st century. This is the new airport control tower, completed in 2019. It features state-of-the-art air traffic control. (AP.)

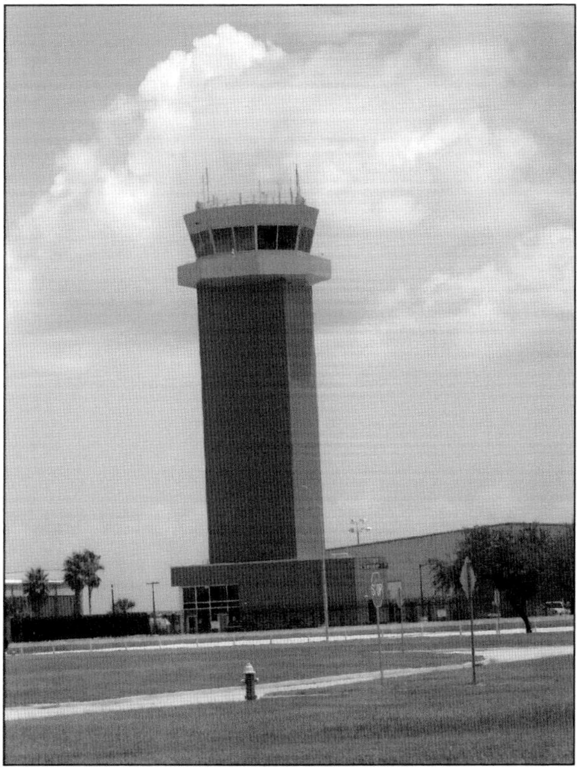

Modern metal hangars for general aviation have replaced structures left over from World War II and the 1950s. These building are certified to withstand hurricane-force winds. (AP.)

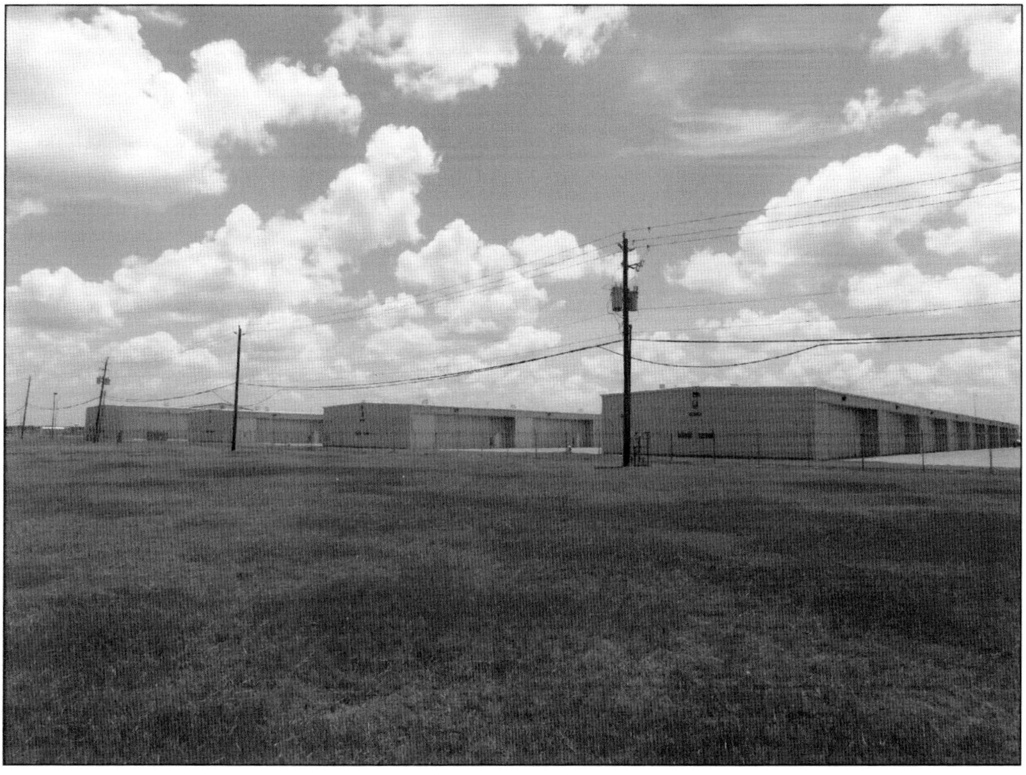

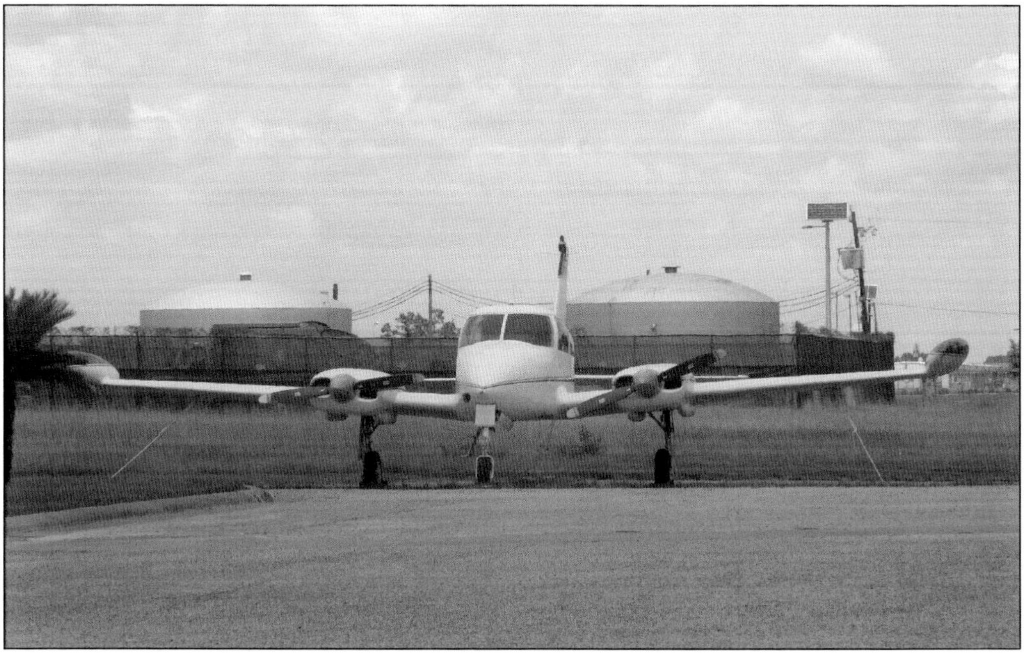

A twin-engine business aircraft is tied down at Ellington. Business twins serve all sorts of functions, from corporate transport to aerial inspection services to charter flights. It is not easily cataloged, and an airport like Ellington provides flexibility airports like Hobby or Bush Intercontinental cannot. (AP.)

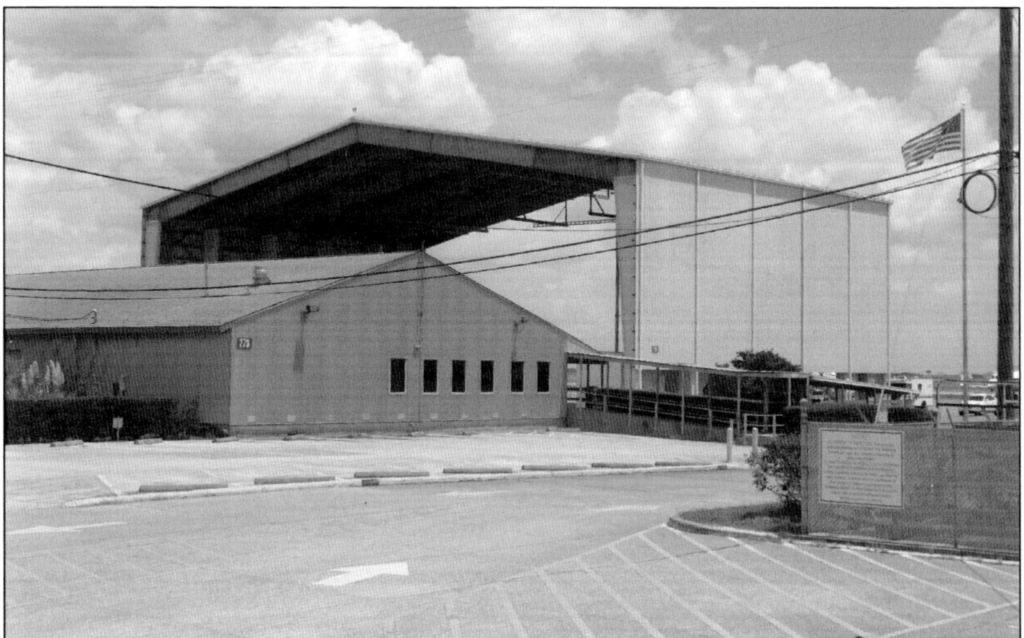

NASA is also at Ellington for the long haul. Perhaps the most visible sign of NASA's presence at Ellington is Building 990, a massive hangar capable of holding the largest aircraft in NASA's fleet. With the ability to roll back the hangar walls at both ends of the building, it offers a distinctive appearance to drivers on State Highway 3 bordering Ellington. (AP.)

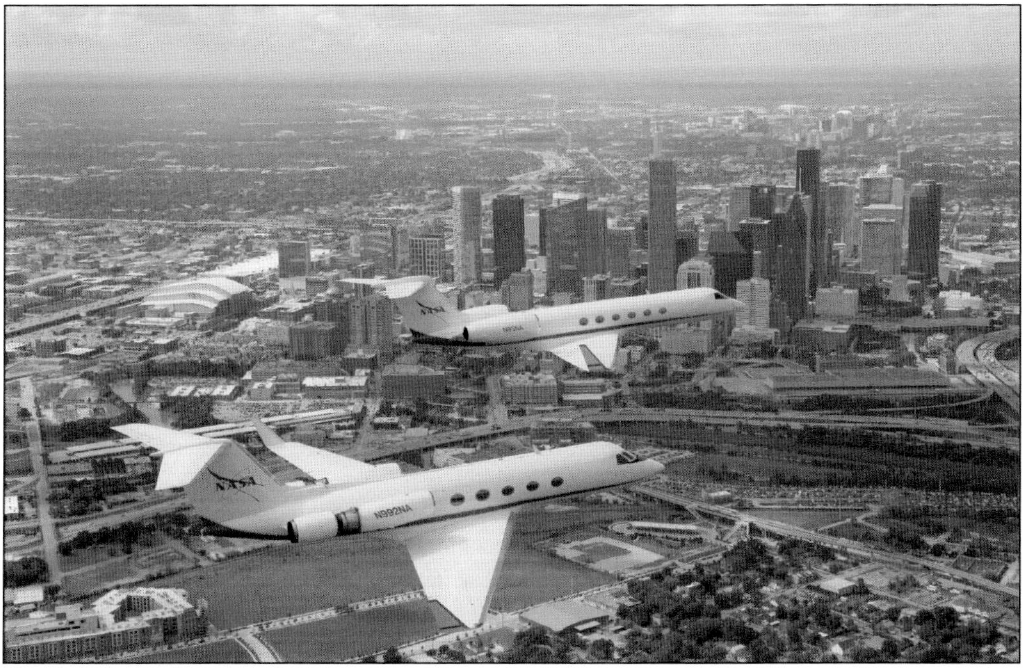

NASA operates a variety of aircraft out of Ellington. While the STA has been retired, JSC operated Grumman Gulfstream jets modified to conduct airborne science. These aircraft can frequently be seen flying out of Ellington or over downtown Houston. (NASA.)

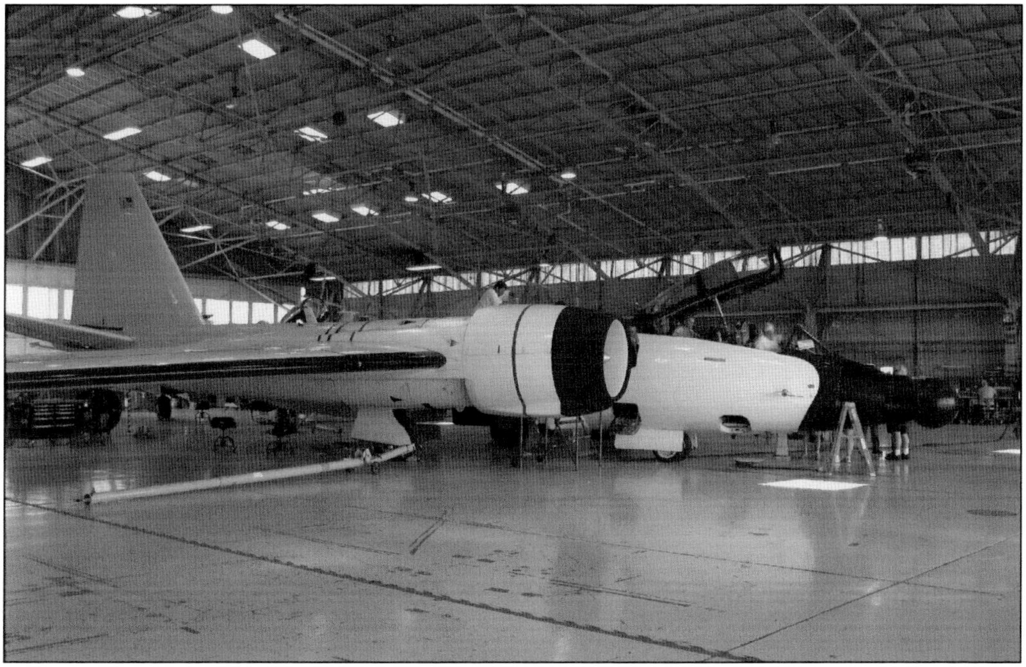

Ellington remains home to two WB-57 aircraft. While designed and originally built in the 1950s, this updated and highly modified version of the 1950s bomber will be conducting high-altitude atmospheric research well into the 21st century. There are no other aircraft that can do the job as well. (NASA.)

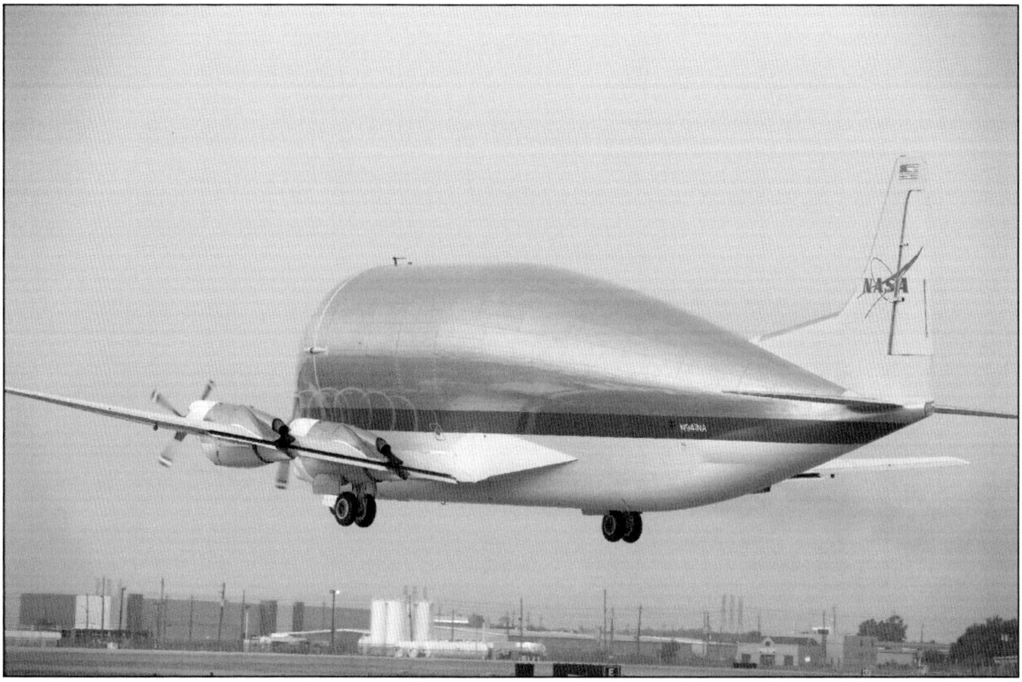

Ellington houses a Super Guppy assigned to JSC. Originally designed to transport Saturn V third stage boosters, the airplane is still useful for transporting large, low-density payloads. Despite its odd appearance, it fills an important function. (NASA.)

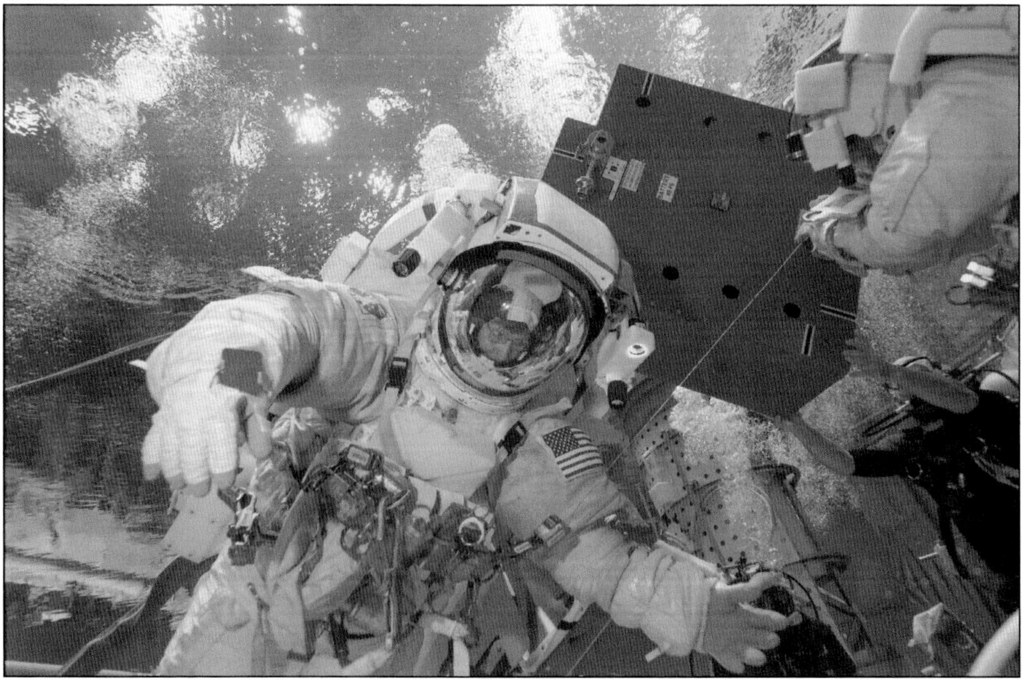

The Sonny Carter Training Facility also remains an integral part of NASA's facilities at Ellington, especially with orbital assembly and maintenance playing an increasingly important role in space activities. (NASA.)

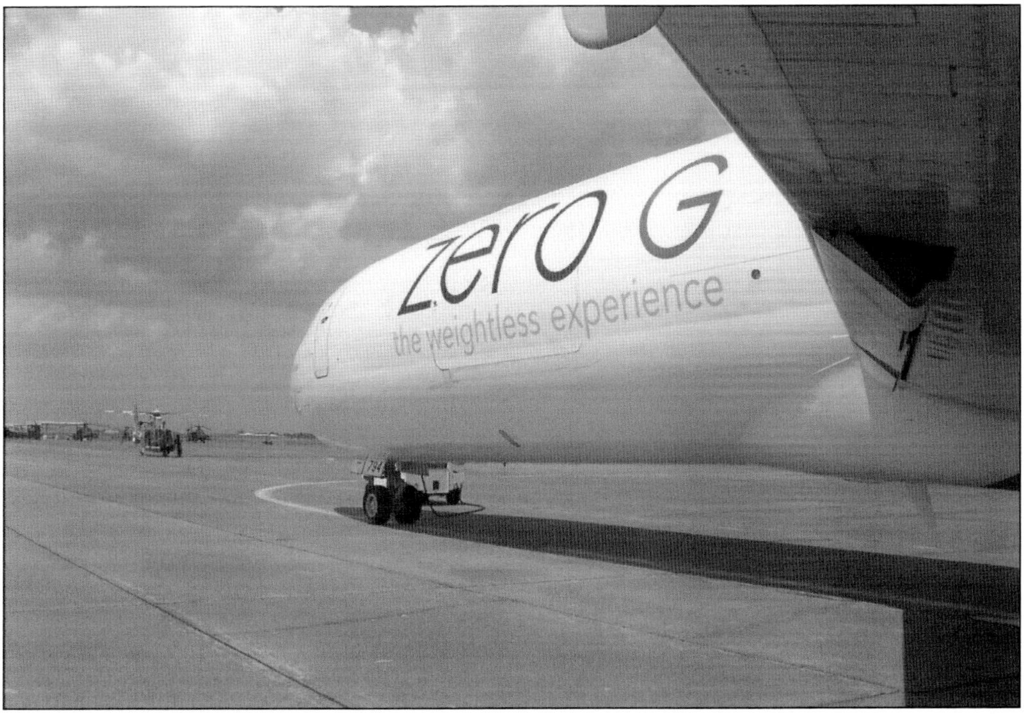

The Zero G aircraft does not call Ellington home, but it visits annually, continuing to offer commercial customers the opportunity to experience zero and microgravity. (DVIDS.)

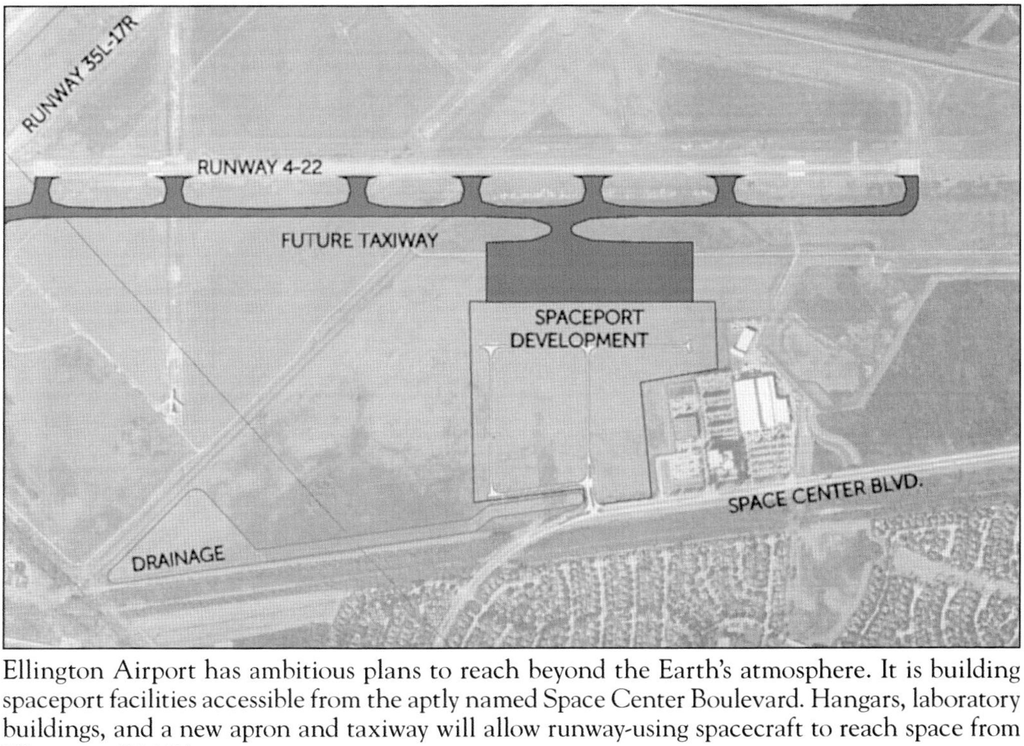

Ellington Airport has ambitious plans to reach beyond the Earth's atmosphere. It is building spaceport facilities accessible from the aptly named Space Center Boulevard. Hangars, laboratory buildings, and a new apron and taxiway will allow runway-using spacecraft to reach space from Ellington. (HAS.)

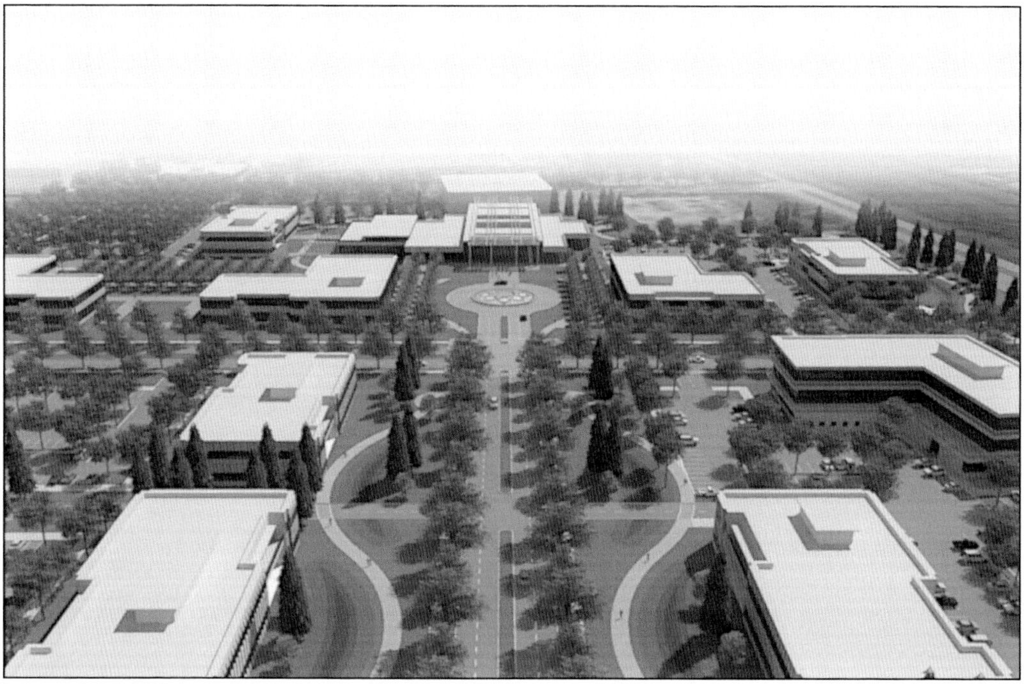

This image offers a view of the Spaceport buildings as they would appear from Space Center Boulevard once the facility is finished. (HAS.)

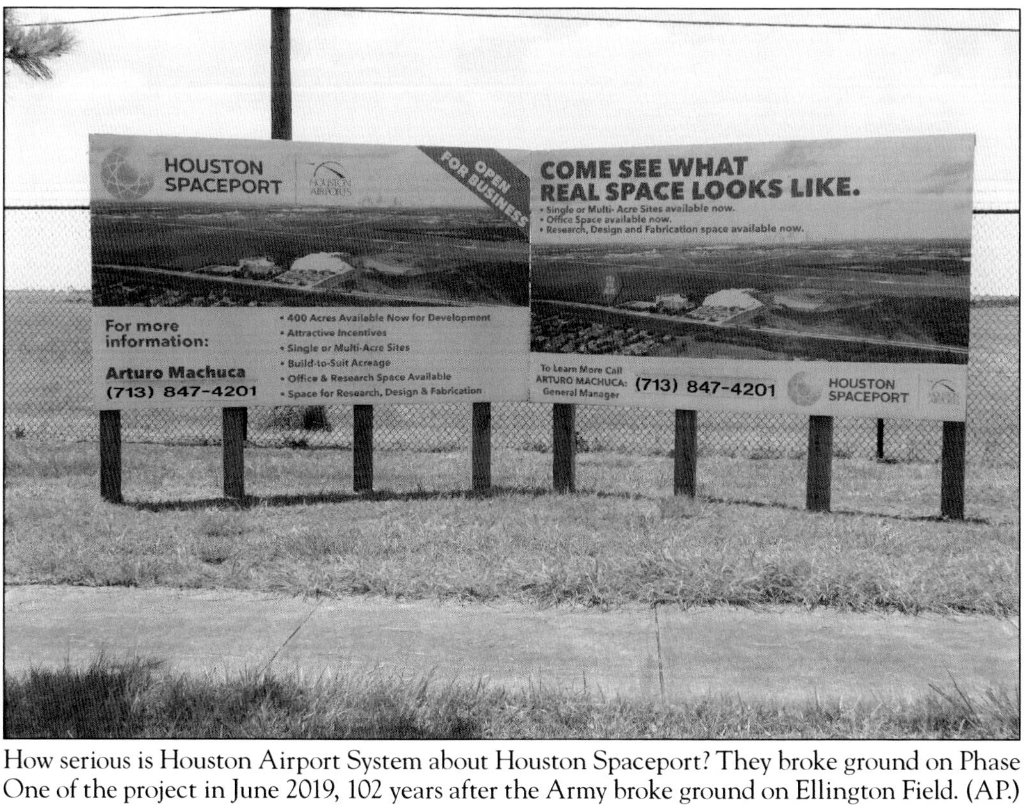

How serious is Houston Airport System about Houston Spaceport? They broke ground on Phase One of the project in June 2019, 102 years after the Army broke ground on Ellington Field. (AP.)

Bibliography

Carlson, Erik. *Ellington Field: A Short History, 1917–1963*. Houston, TX: National Aeronautics and Space Administration, Lyndon B. Johnson Space Center, 1999.

Ellington Air Force Base: Aircraft Observer Training. 3605th Aircraft Observer Training Wing. Baton Rouge, LA: Army & Navy Publishing Company, 1956.

Ellington Field: Houston Texas Army Air Forces Training Command. Baton Rouge, LA: Army & Navy Publishing Company, 1943.

Morrow, Kathryn Black. *Defender of America's Gulf Coast: A History of Ellington Field, Texas, 1917–2007*. Houston, TX: Morrow House Publishing, 2007.

Stack, Lt. Joe. *Ellington, 1918*. Houston, TX: Privately published, 1919.

Westerfeld, Maurice U. *From Jennies to Jets: The Story of the 111th Squadron, 23 June 1923 – 29 June 1973*. Houston, TX: 111th Squadron of the Texas National Guard, 1973.

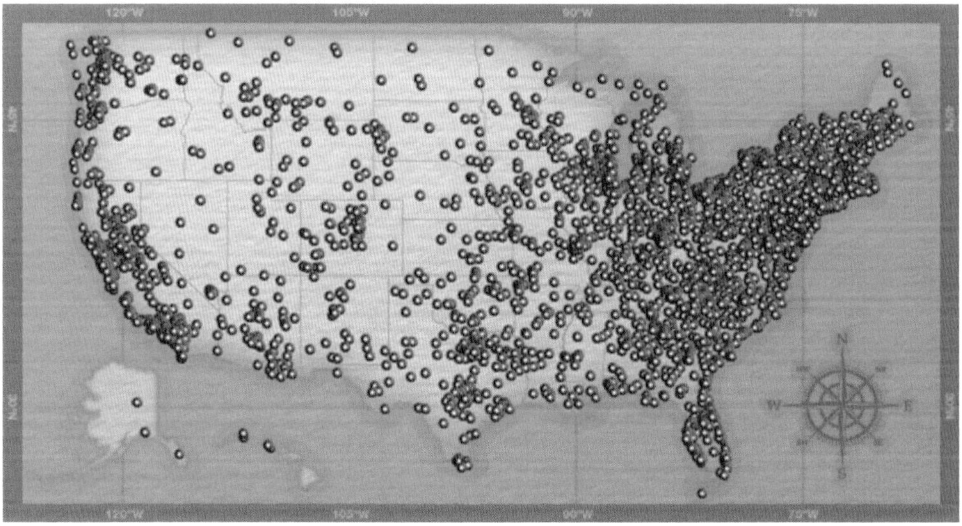